LAWNS INTO MEADOWS

LAWNS INTO MEADOWS

Growing a regenerative landscape

OWEN WORMSER
Illustrated by Kristen Thompson

Stone Pier Press
San Francisco, California

ISBN: 9780998862378

Library of Congress Control Number: 2020937075

Names: Wormser, Owen, author. Thompson, Kristen, illustrator. Ellis, Clare, editor.
Title: Lawns Into Meadows: Growing a regenerative landscape

Printed in the United States of America

Fourth printing: September 2020

23 22 21 20   8 7 6 5

Cover design/illustration by Andy Bridge

Designed and set in type by Abrah Griggs

Contents

THE GENEROSITY OF MEADOWS

I'm checking in on a meadow project of mine. It's late September in western Massachusetts. The sunlight pours through a clear sky with all the vigor of a summer day as I watch a monarch butterfly land on a spike of bluish-purple anise hyssop, and drink from it. For monarchs, which can travel thousands of miles to reach their winter homes, this meadow is like a gas station that lets them refuel on their journey south.

Only a year ago, this one-acre meadow was an empty field filled with old grass and ancient apple trees. Now it feeds a kingdom of pollinators, along with goldfinches, bluebirds, cedar waxwings, chickadees, and many other birds. Mice, voles, deer, and woodchucks forage here. The once-lifeless soil teems with microbial and fungal life that helps it store carbon and makes nutrients available to the plants. That same microscopic world produces sustenance for insects, worms, and other invertebrates, which in turn feed animals like toads and salamanders.

The power of seeds and plants to set in motion and support so much life in so little time is one reason I started my regenerative landscape design practice. I also credit a childhood spent largely out of doors. My parents, inspired by the back-to-the-land movement in the 1970s, raised me and my sister in the woods of rural Maine. By choice, our house lacked most modern conveniences and our nearest, and only, neighbors lived almost a half mile away. Apart from my sister, I grew up without many playmates since I had to ride my bike for several miles to see friends, unless my parents could find time to drive me.

Our daily rhythms revolved around the changing seasons, weather patterns, and the rise and fall of the sun. We used only kerosene lamps, so when the day ended, the night moved into and took over every corner of our house. The only source of heat was the wood my father chopped, so cold days outside meant freezing mornings inside before the woodstoves were fired up again. During the winter months, if you weren't close to one of the stoves or forgot to refill them, the sharp cold was a biting reminder. Trips to the outhouse, always quick, were even speedier.

I got used to the cold, and rarely felt deprived as a boy. I think fondly of the times I sat next to one of those woodstoves at our kitchen table doing homework or reading a book. I liked getting to spend lots of time in the wilderness right outside our door, snowshoeing in winter, tapping maple sugar trees and helping plant our large vegetable garden in spring, and studying the plant and animal life around me all year long.

My parents chose to use as little fossil fuel and plastic as possible both to help the planet, and to become more self-reliant. They weren't purists—we had a car, I went to public school and later boarding school, we used dentists and doctors, and we bought clothes rather than making them. Still, they made many sacrifices in pursuit of a simpler life and a stronger connection to the earth. Traveling in the winter wasn't an option because someone had to be at home to keep the stoves going. And in the summer my parents were generally too busy preparing for winter, by growing our food and chopping wood, to do much of anything else.

I'm grateful I had the chance to grow up living simply and close to nature. But as an adult, I really appreciate not having to heat up water before taking a shower or drive to a laundromat to wash my clothes. I'm very glad I don't have to get up in the middle of a freezing cold night to use the outhouse. And I relish every time I get to watch a good movie or television show well after the sun goes down.

You don't have to live off the grid to help the environment. There is a reasonable middle road to lightening your burden on this planet. Among the *many* options easier than giving up everything to live in a cabin in the north woods of Maine: Consume less. Buy locally. Cut back on meat and dairy. Compost your food waste. Grow some of your own food. Use public transportation whenever possible. Fly less. You can also grow a meadow instead of a lawn.

Lawns have become something of a national obsession. We waste an enormous amount of resources every year maintaining a closely cropped area of turf that totals more than 63,000 square miles, about the size of Washington State. By another measure, over forty million acres of land in the continental United States were found to have some form of lawn on it. This massive footprint makes lawns the biggest irrigated crop grown in the United States, and it sucks up an outsized amount of fossil fuels, fertilizer, chemicals, and water. Landscape irrigation is estimated to account for nearly one-third of all residential water use, totalling nearly nine billion gallons per day or almost 13,500 Olympic-sized swimming pools worth of water.

To be fair, lawns serve a purpose. They offer places to play, which is something I was glad for as a kid. My parents may have grown much of our food, but they also made sure we had enough lawn to kick a soccer ball around on. And lawns have other uses beyond sports: As a designer, I sometimes use strips of turf as pathways in gardens or to frame a perennial meadow. But the vast majority of mowed lawns offer none of these advantages. Instead, they're a danger to the planet and to many living things, including your children and pets.

A meadow is what can happen when you give the earth a chance to heal itself. When planted properly, it fills out easily and grows almost entirely on its own. With every year in the ground, meadow plants support more life and build healthier soil. This makes them quite efficient at parking carbon—just the opposite of a resource-guzzling lawn. Lawns are among the ways we burden nature. Meadows are far more generous, giving back to the earth much more than they take.

I decided to write this book because, along with so many people, I'm alarmed by how quickly our planet is overheating. Farmers are on the frontlines of global warming and forced to deal with a longer growing season, and more flooding, drought, and extreme storms, as well as new batches of pests swarming northward as the country warms.

Meadow builders have it much easier. While I've noticed more pests, and more weeds, too, a meadow is so resilient it can put up with just about anything. The wide variety of plants in a meadow offers some protection. But the plants themselves are strong, too. Many native plants across the country are struggling to survive climate-induced weather extremes. Native meadow plants? Not so much. And yet too few of us know that planting a meadow is even an option.

In this book, I'll outline steps you can take to create your own regenerative landscape, one that improves the environment by increasing biodiversity, enriching soil health, and removing carbon dioxide from the atmosphere. I'll help you evaluate the lot, yard, or community space you want to turn into a meadow to see if it's suitable. (If it gets enough sun, it probably is.) You'll find guidance on how to design your meadow, and a list of twenty-one easy-to-grow perennials you can use as a starting point. This book also includes tips for introducing a meadow into your neighborhood when everyone else has a well-groomed lawn.

Recently, a friend of mine did something very simple. She allowed common milkweed to grow in one of her garden beds instead of pulling it out. Milkweed attracts monarch butterflies and their caterpillars—in this case, lots of them. Later that season we counted no fewer than fifty monarch chrysalises that had hatched. When I visited my one-acre meadow one last time before winter, I got to see a few more of these beautiful creatures sipping their last drink of nectar before flying off to Mexico. If more of us go ahead and plant meadows, they'll have a much easier time finding their way back.

LAWN TROUBLE

Not long ago I visited a young couple who'd bought a house in Northampton, Massachusetts, and wanted me to come look at their lawn. Walking up their driveway, my eye was drawn to a lilac bush near the house covered in radiant blue-purple flowers, and I picked up the sweet fragrance perfuming the chilly spring air. Other than that, there was nothing much to see. The entire half-acre yard was all lawn, or used to be. Most of the grass was gone, leaving remnants of turf competing with dense clusters of short, scraggly weeds. Apart from some insect activity around the lilac blossoms, the yard was still.

I've always avoided renovating or installing lawns. Instead, I usually make the case for getting rid of them entirely. Sometimes, circumstances make it easier. This is especially true when a lawn collapses, which is what happened here, and something that happens often when a house changes hands. During the transition period, lawns that have been sustained by chemical fertilizers, watering, and pesticide treatments are often ignored. Once their life support is removed, these lawns all too easily fail.

The homeowners asked me if there was anything I could do to revive the lawn. But when I told them they had other landscaping options, they wanted to hear more. As with so many people, they didn't realize there even were alternatives. All they needed to hear was how low-maintenance a meadow would be—they already had their hands full with the new house—and they were convinced.

Before the end of the growing season, we had turned over that sad, exhausted lawn and seeded in a meadow. The space eventually filled with long, waving grasses, bright yellow, orange, pink, and purple flowers, and a crowd of butterflies and bees to help keep that lilac company.

WASTELAND

A fresh green, perfect-looking lawn is not as healthy as it appears. It's locked in a cycle of chemical dependence that weakens it considerably, making it vulnerable to disease and death once the support stops. When grass is given topical fertilizers, its roots remain shallow since there's no incentive to push deep into the earth on a quest for more nutrients. Shallow roots have a harder time reaching moisture underground, which makes the grass overly dependent on regular waterings.

Chemically based treatments further damage a lawn by harming the healthy microbial and fungal life that helps keep plants well nourished. Without the support of these organisms, conventional lawns can quickly become biological wastelands, in spite of their green good looks. The lack of living soil also makes it harder for the ground to absorb water when it rains, leading to higher levels of stormwater runoff and erosion.

The problem with lawns doesn't end at their neatly clipped borders. Lawns significantly degrade the ecological health of land and water, and contribute to global warming. Take synthetic fertilizer, which is largely responsible for keeping those short-rooted grasses green. It's estimated that Americans use one hundred million tons of fertilizer on their lawns each year. For every ton of fertilizers manufactured, two tons of carbon dioxide are produced.

Since plants can't absorb all that fertilizer, some of it evaporates in the form of nitrous oxide, a greenhouse gas about 300 times more potent than carbon dioxide. Increased fertilizer use over the past fifty years has been linked to a dramatic rise in atmospheric nitrous oxide. The rest of the excess washes out of the soil, along with herbicides and pesticides, and pollutes lakes, streams, and groundwater. Conventional agriculture is infamous for producing enormous aquatic dead zones from high levels of chemical runoff. Far less well known is that homeowners use up to ten times more chemicals per acre than farmers do, according to the US Fish and Wildlife Service.

One significant source of chemical input is pesticides, which are not safe, in spite of how commonly they're used on lawns and to grow food. In fact, they have to be registered with the Environmental Protection Agency (EPA) because they can be so dangerous. But pesticides aren't always rigorously tested before they're made available to the public.

Thanks to a loophole in EPA regulations, the agency can "conditionally approve" pesticides based on data provided solely by the manufacturer. In 2013, the Natural Resources Defense Council (NRDC) found that close to 65 percent of the 16,000-plus pesticides approved for agricultural and consumer use took advantage of this loophole. Today, the agency continues to greenlight chemicals before a full evaluation of health and environmental impacts has been completed, though it has said it's working to make testing more rigorous.

The lack of consistent oversight over the course of decades is one reason the pesticide industry has a history of releasing products that are a grave danger to plant ecosystems and life itself, with dichlorodiphenyltrichloroethane (DDT) among the best-known. A class of insecticides called neonicotinoids, or neonics, which attack parts of insect nerve cells that are similar to those found in humans, was supposed to be safer. The EPA went ahead and approved its use before fully understanding the terrible impact it would have on monarch butterflies, bees, and birds.

Since their introduction in the 1990s, neonicotinoids have become the most widely used class of insecticide in the world, and the numbers of bees and other pollinators have plummeted. The total biomass of insects worldwide is down by a precipitous 80 percent, and because most birds eat insects, the populations of more than 75 percent of songbirds and other birds that rely on agricultural habitat have also significantly declined. Europe banned neonicotinoids for agricultural use in 2018 because of their impact on pollinators and birds, but the EPA is still telling Americans that many of them are safe.

Herbicides are no better. In one prominent example, the World Health Organization found sufficient evidence of carcinogenicity to classify glyphosate in 2015 as "probably carcinogenic to humans." Glyphosate is the active ingredient in Roundup, made by Monsanto (now known by its parent company name, Bayer) and the most popular weed killer in the country. In 2007, when 180 million pounds of it were being used nationally, glyphosate was found in 86 percent of air samples and 77 percent of rain samples taken in Mississippi by environmental scientists. In 2016 alone, more than 286 million pounds of it were used.

Instead of creating a culture of accountability to compensate for these transgressions, the pesticide and herbicide industry spends vast amounts of money trying to convince the public their products are safe. So far these tactics have largely worked, though at least some of the science presented by corporations is being exposed as unreliable. Bayer has lost multiple lawsuits brought by plaintiffs who claimed the company failed to warn them of the increased risk of developing non-Hodgkin's lymphoma and other forms of cancer while using the weed killer. As of this writing, it's dealing with close to 100,000 similar lawsuits. In spite of this, the EPA recently reapproved the use of glyphosate.

This makes it likely people will continue to use these chemicals, however harmful. Lawn care chemicals account for the majority of wildlife poisonings reported to the EPA. Pets kept in yards treated with chemicals experience significantly higher occurrences of malignant lymphoma—70 percent higher—than pets in untreated yards. Repeated use can also degrade soil biology. With all this in mind, it's clear these chemicals are probably not something you want in your yard around your pets, much less your children.

It's a lot to think about, and clients willing to hear me out on the dangers of tending a bright green space often ask me about organic lawns. Aren't those better?

Yes, and no. Organically maintained lawns do not rely on herbicides or pesticides. If treated with a good slow-release organic fertilizer, or compost, they can even support healthy microbial life underground. But most lawns need many gallons of water to thrive. And then, of course, there is mowing.

MOW, MOW, MOW YOUR LAWN

Mowing lawns happens to be a major source of pollution. The EPA estimated that seventeen million gallons of gasoline are spilled every year in this country just while filling lawn mowers—far more than the eleven million gallons spilled by the Exxon Valdez supertanker.

After they're gassed up and running, lawn mowers emit ten times more hydrocarbons than a typical car for every hour of operation. This means that in one hour, a commercial lawn mower, the kind used by professional lawn care companies, spews as much smog-forming pollution as driving a 2017 Toyota Camry for 300 miles. Brand-new gas-powered residential mowers running for just one hour are not much of an improvement; the amount of emissions released is the equivalent of driving a car for one hundred miles.

The trouble with lawns doesn't stop there. Greenhouse gas emissions from mowing, along with fertilizer and pesticide production, watering, leaf blowing, and other lawn management practices, were found by a University of California-Irvine study to be four times greater than the amount of carbon stored by grass. In other words, a conventional lawn produces much more carbon dioxide than it absorbs.

When looked at objectively, it's obvious lawns are an ecological catastrophe. But in addition to the environmental costs, they also demand a lot of our time and money. The average lawn owner spends 150 hours a year tending to their grass, and collectively we spend forty billion dollars on lawns annually.

So, how did we get here? How did close-cut greenscapes become our number one crop?

Before the mower rolled onto yards in mid-nineteenth-century England, lawns were far too labor-intensive for the average person to maintain. Managing cropped grass was possible only if you could afford to pay people (lots of people) to trim it by hand. In 1830, British engineer Edwin Beard Budding started to change all that with the invention of his reel mower, which was inspired by a cloth-cutting reel in a local factory. But early models were cast-iron, heavy, and unwieldy. As soon as the original patent expired, others rushed in to improve on it.

In 1870, an inventor named Elwood McGuire of Richmond, Indiana came up with a lightweight version. Suddenly, yards could be mowed relatively effortlessly in mere hours. It didn't take long for modern sports like football, baseball, and soccer to evolve along with the sleek new landscapes, and soon people everywhere started planting their own. To meet the demand, gasoline-powered mowers began to appear by the end of the nineteenth century. But it wasn't until after World War II that cheap, mass-produced mowers became widely available.

The gas mower offered an opportunity for every homeowner to reenact the American dream of Manifest Destiny. It was a chance to subdue and control the wilderness on their own terms, on their own plots of land. While this mindset may not have been conscious, it was enthusiastically embraced. In 1946, just 139,000 gas-powered rotary mowers were sold. By 1959, with lawns having become a defining feature of the American landscape, sales would hit 4.2 million—and they haven't slowed since. It's estimated that well over 5 million gas powered mowers are sold in the US every year.

REGENERATIVE SCAPES

As the child of organically inclined parents who believed in gardening for pleasure and sustenance, my interest in natural landscapes started early. Inspired by Maine homesteaders Helen and Scott Nearing, the authors of the seminal book *Living the Good Life: How to Live Sanely and Simply in a Troubled World,* my parents grew and very much relied on a large vegetable garden to help keep us fed. My mother also loved perennial flowers and cultivated a variety of them, just for the fun of it. Some of those plants, like *Coreopsis* and *Liatris*, were my first introduction to meadow plants, and are still very much in my life.

My parents taught me you don't need synthetic chemicals to grow healthy food and flowers. From them I learned that the success of a garden begins and ends with the quality of your soil. We turned the food scraps and garden waste we collected into compost, and used it to cultivate the dark, loamy soil in which we grew our carrots, broccoli, peas, and beans. I remember the smell of it; the rich, soft soil exuded an aroma that made me feel like the earth itself was exhaling.

By the time I graduated from high school I knew I wanted to spend my life outside, growing gardens and landscapes. Setting up a landscape design business shortly after college seemed like a logical next step. I planned landscapes that fit in with the surrounding environment, and don't need watering, fertilizers, or herbicides once established. It didn't take long to realize that if I truly wanted to build sustainable landscapes, I'd have to learn how to grow a meadow.

Back in the late 1990s, when I started my practice, most people had no interest in getting rid of their green grass. When the topic of lawns came up in client meetings, it was usually a request to install one, which I'd decline. I remember how frustrating it was not to be able to point out a meadow—there just weren't any local examples. If people had seen the one I visited on trips to Storm King Art Center in Cornwall, New York, where it grew among the large sculptures, or had a chance to take in the simple beauty of a hayfield in full bloom, they'd probably have considered planting a meadow of their own.

I didn't have a meadow to show them, but I could listen to their concerns. Most people I met complained about the time and money involved in maintaining a lawn, so I told them established meadows need none of the watering, chemicals, or fossil fuels that lawns do. All they need to thrive, I said, is a half day of sunshine. To those who were tired of their lawn, but not sure they were ready to part with it, I suggested they keep some turf for playspace or pathways. Preserving a little lawn won't interrupt the integrity of your meadow, nor its ability to serve as a hangout for butterflies and ladybugs.

In those early days of talking up meadows, I won over very few converts. My meadow-building efforts only gained traction after I'd managed to actually plant a few. Most people had to see for themselves the year-round multi-hued grace of a meadow, and its lively population of wild things.

These days people still like hearing about the low-maintenance demands of a meadow. But I'm finding an audience increasingly receptive to a meadow's regenerative powers, which are all the more compelling when compared to the turf that meadows can replace.

HOW MEADOWS STORE CARBON

About half of all carbon released into the atmosphere every year is absorbed by the planet's oceans, plants, and soil. While most studies focus on the carbon-storing powers of oceans and forests, the research on grasslands finds they're a cost-effective and scalable solution for carbon absorption. Given how long it takes for trees to grow, meadows can perform as well as or better than a forest when it comes to sequestering carbon underground.

The Carbon Cycle

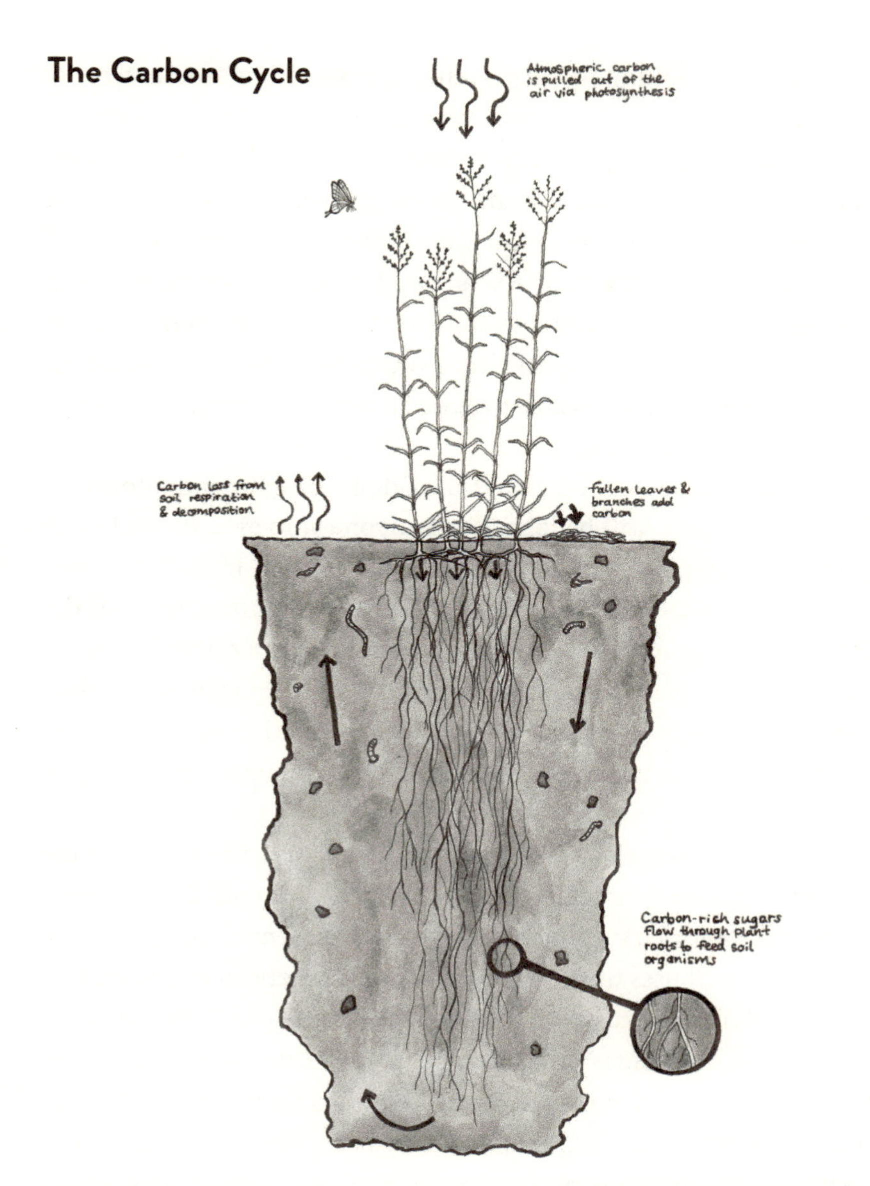

Plants help the soil store carbon through the process of photosynthesis. Plants absorb carbon dioxide from the atmosphere. They exhale the oxygen but retain the carbon molecule, which they use to create the starches and sugars they feed on. Through their roots, they share some of that carbon-based food with organisms living in the soil, such as mycelium and microbes. Those organisms absorb the carbon and when they die, it remains in the soil.

Grasslands hold about 20 percent of global carbon stocks, according to Project Drawdown, a nonprofit that assesses climate solutions. In one study, North American grasslands were found to store between four and a half and forty tons of carbon per acre—and that's just in the top twenty centimeters of soil. In another, the soil under two and a half acres of healthy prairie was found to have absorbed as much carbon as what 150 cars emit over the course of a year. Perhaps most importantly for anyone considering swapping a lawn for a meadow, an established meadow is able to store 70 percent more carbon than a monocrop, like turfgrass.

The way meadows pull carbon dioxide out of the air and store carbon underground will be familiar to anyone who ever studied basic plant biology: photosynthesis. This is the process by which plants absorb carbon dioxide and sunlight to create their own food, in the form of glucose and other sugars. Oxygen is a byproduct of this process. After the plant exhales oxygen back into the air, it synthesizes sugars using the carbon and combining it with other molecules, like hydrogen taken from water. These carbon-based sugars are what give plants the energy to grow.

Over the course of a meadow plant's life, a significant amount of the carbon it converts from a gaseous to a solid state ends up in the soil. Some of those carbon-rich sugars created by a plant are sent to its root system, where the plant uses them to attract and feed beneficial microbes, bacteria, and fungi. In exchange, those organisms make available nutrients that nourish the plants. When the organisms die, they leave behind carbon, which remains in the soil for as long as it's undisturbed. This carbon cycle explains how a meadow, with its many and varied deep-rooted perennials, becomes a very effective carbon sink.

Meadows won't ask much of you on their way to becoming eco-superstars. Perennial meadow grasses and flowers are deep-rooted and opportunistic, which means they can adapt to poor soil and tough growing conditions. Without much input, if any, they can reclaim tired, near-ruined soil by adding organic matter in the form of decomposing plants and soil organisms. The ability to make do in less-than-perfect conditions makes their soil-building powers even more impressive, and useful.

We love our lawns, but apart from the most unapologetic and passionate lawn enthusiasts, I'm convinced many of us could easily fall in love with meadows given the chance to stand in one, or watch one grow. If we do, we'll see the dried-out look of a late summer lawn replaced by a richly textured, ever-shifting colorscape. And we'll hear the roar of lawn mowers fall silent so the sound of chirping birds and crickets can rise up and be heard, and we all get to breathe a little easier.

FIELD STUDY

I first visited the Eric Carle Museum of Picture Book Art in Amherst, Massachusetts, with an assignment: to design a meadow right outside its doors. Author and illustrator Eric Carle was hoping to create an ecologically friendly, living memorial to Bobbie Carle, his wife and the museum's cofounder, who had recently passed away. A landscape architect had designed the cement walkway that looped through the one-acre site, which was essentially a dry lawn dotted with aging apple trees. But aside from the path and the trees, the space looked empty. No children, and no insects, birds, or squirrels either.

As I stood in the shade of one of the more stately trees, I thought about how to turn this unappealing site into a draw for museum visitors, one that could live up to Eric Carle's dream. I started by studying the one hundred-year-old orchard, beginning with the tree I was standing under. It had a full, green canopy and leaves unblemished by disease and insects. The other fifty or so apple trees ranged in appearance from fully thriving, like this one, to gaunt stragglers on their way out. A decent number had already died and been removed.

As I walked among the trees, the ground beneath me was hard; it felt almost as jarring as walking on pavement and far from the spongy feel of richer, healthier soils. When I kicked at the sod, it sprayed everywhere, indicating weak shallow roots. Its clay-like appearance and texture wasn't all that different from adobe, though not nearly as dense.

The poor quality of the soil didn't worry me too much. After years of witnessing the adaptability of meadows I knew that certain species, like yarrow, butterfly weed, and purple coneflower, could grow just fine in it. Still, the conditions in that field didn't exactly inspire confidence. Worse, I knew that most commercial orchards dating to the first half of the twentieth century had been sprayed regularly with lead arsenate, a compound once considered safe that can linger in soil even one hundred years later. I wondered how it had affected the health of underground organisms, or if it would stunt the flowers and grasses I wanted to plant.

I continued peering at the ground as I walked the rest of the site. An occasional grasshopper or cricket scrambled away from me, but in general there was a notable lack of insect life—other than the ants. Thousands of them, their little finely grained homes rising out of the many bare spots surrounding me. If so many ants could dig into the soil, it probably wasn't as dense as it appeared.

On any new meadow project it's tempting to jump ahead in the design process and start thinking about the flowers and grasses you want to plant. This assignment was no exception. Almost as soon as I arrived, I started picturing the colors, textures, and height of the plants I'd like to see in that field. But without first assessing and fully understanding the conditions of your site, you will, for sure, end up picking the wrong plants.

At The Carle, the tired, hard-packed soil was just one of the challenges I'd have to work through. Another was to preserve as many apple trees as possible, per the museum's request. Apple trees' roots grow close to the surface. This is only a problem if you've decided not to use herbicides, as the museum and I had. (Not to mention it's never a good idea to spray herbicides where children will play.)

Clearing a big stretch of grass without using chemicals generally calls for a rototiller. But tilling, or turning under, the sod would very likely damage the shallow tree roots. This meant we had to figure out how to install a meadow into the existing lawn rather than clear the site, which is the preferred way to prep a meadow. And it had to happen quickly.

The construction of the walkway had run into delays, which pushed back our start date. It was already June, with the meadow's public unveiling slated for early October. There was no way we'd have a meadow established by then, but I wanted it to at least look like it was well underway. I could have sped along the process by planting tiny young plants, known as plugs. But the space was far too large to fill with live plants, so the only option we had was to seed it.

There are two well-known rules for establishing a meadow. One is to seed into bare ground. The other is to plant native perennial seeds when it's cool outside. To get the meadow up and running quickly enough to look full and beautiful by the next summer, I'd have to break both rules.

I had no experience in this area so I reached out to three fellow professionals to get their take. None of them could help, and the books and meadow-related websites I consulted were no better. Seeding into turf makes it harder for seeds to make contact with soil, plus they have to put up with competition from other plants. And laying down seed in the heat of summer can delay growth since, for many seeds, germination is triggered by cooler weather.

So I decided to play the odds the only way I knew how—by carefully choosing species that promised the best chance of thriving on this particular site. I chose grasses like little bluestem and nodding fescue, which are hardy enough to find nourishment even in dense soil. I picked a few shade-tolerant flowers, including Solomon's plume and meadow rue, to plant under the shade of the apple trees. To meet the demands of a meadow on a tight schedule, I chose a range of fast-growing seeds, like wild bergamot and anise hyssop. Since I was also planning for the long haul, I chose plants that like dense, often very dry, low-quality soil, including various types of aster, beardtongue, rose milkweed, and a native bellflower.

When it came time to plant, we relied on a drill seeder to help penetrate the existing lawn and give our seeds the best shot at growing quickly. The machine slices rows of small V openings in turf, drops seeds into each one, and covers them up. For seeds too small to fit in the drill seeder, we used a hand spreader, a hopper with a simple hand crank system that sprays seed evenly as you walk.

Even after all the planning, I steeled myself for the possibility that few, if any, of the seeds would germinate and work their way up through the sod. At best, I told myself, it would take a year from the time it was planted for this meadow to grow into the expanse of lush, waist-high grasses punctuated by the bright yellow, blue, and white blossoms that I envisioned.

About six weeks after the planting, my skepticism seemed well-founded. Seedlings were nowhere to be found. Instead there were weeds, lots of them, especially along the dismal strip of exposed soil left over from the walkway construction. My crew and I cut them back with a weed wacker and hoped for the best.

Two weeks later, I stopped by again, and continued to do so every chance I got. Every time I looked for meadow plants, and every time I found weeds. It's not unusual for weeds to sprout after planting a meadow. But an uncleared site is much more vulnerable to being taken over. We kept on top of the new growth, eventually switching to a scythe—more pleasant than a loud, smelly weed wacker and, in this case, more effective. And I tried to keep my anxiety in check.

October came and went. At the opening ceremony, the meadow looked pretty nondescript, even unkempt. It was certainly not attractive. The lawn grasses I'd hoped would have shot up by then hadn't grown much at all. I reminded museum officials we'd gotten a late start and that the real show would be next year, when the perennial grasses and flowers had a chance to take root. I hoped I was right. I was dealing with factors I'd never encountered before and wasn't all that sure myself.

I resumed my watch the following March. The first thing I saw when I arrived was a tangle of dried-out weeds and scruffy lawn grass, about what you'd expect after a long winter. Almost two months later, the old orchard blossomed into overarching sprays of delicate white flowers. But the site still didn't look very meadowy. In one panic-stricken moment, I considered asking for a do-over and reintroducing more seed. Then I thought again about the resilience of the seeds I'd chosen. So I decided to just wait and see—even if it meant putting up with a few more sleepless nights.

Sure enough, one year after the planting, the meadow grasses began filling in. I'd missed seeing the tiny seedlings but they must have been there all along, taking root among the weeds and old grass. A few weeks later, the perennial flowers began to appear. Wild bergamot were the first to show up, the pale pink tubules drawing in pollinators of all kinds that buzzed between the blooms, and they were quickly joined by five other species of flowers. I learned that foxes and deer had been spotted roaming among the grasses. Along a treeline on the meadow's edge, hawks had taken up fresh postings.

Every time I stand in that meadow, a part of me marvels that all this grew from seeds that barely filled half a five-gallon bucket. Not only was it an enormous relief to see those grasses and flowers finally push through the matted turf, it affirmed the impressive strength of native meadow plants. In the end, all I had to do was allow the site conditions to drive my plant choices, be patient, and beat back all those weeds.

SIZE UP YOUR SITE

My first dates with prospective meadows usually take place in yards with tired lawns. But man-made meadows can also spring from city lots scattered with trash and cement rubble, fields riddled with brush and weeds, and wide strips of mowed turf lining busy roads. None of these conditions disqualify a site. The single most important requirement is pretty straightforward: If your plot gets at least a half day of direct sunlight, you can grow a meadow. But there's more to it than that.

A meadow builder's goal is to create a landscape as ecologically stable as a wild meadow so that, once established, nature can take over. Location is important, of course. Land hemmed in by neighbors partial to well-trimmed lawns, for instance, may call for framing your meadow with neatly cut grass. Steep slopes can be harder to weed and tend to, so ask yourself if you're up for that. If your site is next to a busy road, you might want extra-tall grasses to help screen it. And so on.

But the nitty gritty of a good site evaluation takes place on the ground level, sussing out how moist your land is and the quality of your soil. In putting together a site profile that points the way to a compatible set of plants, I suggest taking the following steps.

Find your growing zone. Climate change may be altering weather patterns around the globe, but broad warming and cooling trends still mostly hold. To find your growing zone, consult the hardiness zone map produced by the US Department of Agriculture (USDA), which is used by most growers. It focuses on the coldest winter temperatures in your region, which play a big role in determining which plants can thrive there.

Note: The government produces *two* hardiness zone maps—the USDA map and another managed by the National Oceanographic and Atmospheric Administration (NOAA). The USDA uses average temperatures from 1976 to 2005 and the NOAA uses average temperatures from 1981 to 2010. The maps match for the most part but there are differences. The USDA's map is geared toward agriculture and helping people produce a healthy crop, while the NOAA map tracks climate change more closely.

Study what's already growing. The plants already living on your site are pretty good indicators of the quality of your soil and moisture levels. Jewelweed, a common annual weed with attractive orange or yellow flowers, only grows in areas that never completely dry out, for example, so its presence is a sign of very moist, and even wet, conditions. *Atennaria* or catsfoot, a weed species that often grows in lawns, only thrives in very dry, sandy soil. Dandelions indicate that your yard gets plenty of direct sunshine. If moss is growing in between

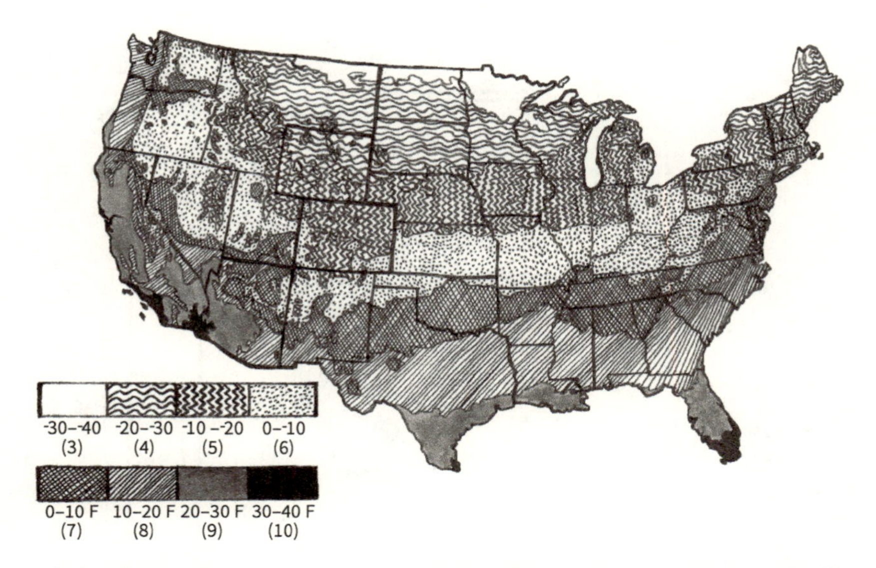

The hardiness zone map is what growers use to determine which plants are most likely to thrive in a given location. North America is divided into eleven hardiness zones based on the average annual minimum winter temperature in each region. Eight zones are in the contiguous United States. The northernmost zone is 3; the southernmost is 10. Source: USDA.

blades of grass it means those areas are too shaded for meadow plants to prosper.

Other clues: if your weeds are robust, with full, green leaves, your soil is probably rich. The same goes for lawn grass that grows a deep green without fertilizer. If, on the other hand, the weeds and grass in your yard are stunted with thin leaves, it's likely you have poor soil, which isn't necessarily a problem since many meadow plants do just fine in it.

To figure out what your weeds are telling you, I recommend getting a field guide to local weeds or downloading a plant-identifying app or site, like iNaturalist or GoBotany.

Check how well your soil absorbs water. When I first visited the site that would become The Carle meadow, it was dry, hard-packed, and

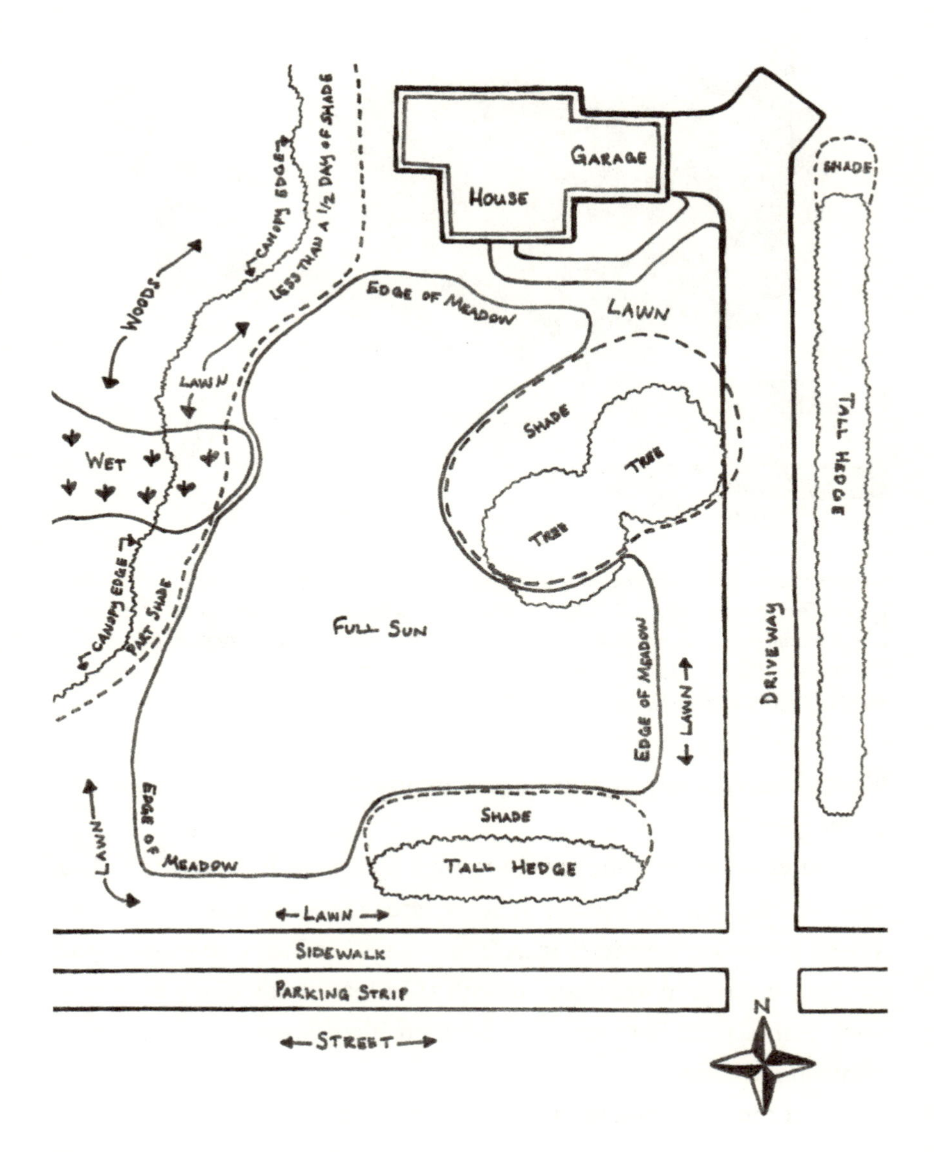

When figuring out where to locate a meadow it can be helpful to conceptualize it on paper. You want to situate your meadow where it's likely to get the most sun. This map also shows how a mowed strip around the perimeter can nicely frame it.

dusty, thanks to the construction involved in installing the concrete path. It was also during a summer that hadn't seen much rain.

When I dropped by again almost two months later, we had just come through a heavy downpour. All that water created many small, shallow pools under the apple trees, turning the exposed soil from a light brown, dusty consistency into dark brown mud. I returned a couple of days later to find that those puddles had drained quickly and the remaining muddy patches dried up and blended into the rest of the soil.

The evenness of the soil was good to see. When a single plot contains more than one distinctly different zone, say wet and dry areas, it makes sense to choose different plants for each and scatter the two seed blends in their respective areas. If conditions are fairly uniform, which has been the case for most of the meadows I've worked on, you only have to choose one set of plants.

A few of the signs indicating your soil is prone to being wet or dry:

- *The site dries out entirely after the end of spring rains.* This is typical of arid zones and, no surprise, you'll need dry-weather plants.
- *Water tends to pool after it rains.* Standing water for more than a day or so indicates poorly draining soil, and this can be a problem. For any meadow to do well, the soil shouldn't sit in water for very long.
- *Water wells up from underground, especially in the spring.* You'll need plants that can handle the moisture.
- *There is ledge rock or stone near the surface.* Bedrock can cause ground to dry out fairly quickly, so drought-tolerant plants may be a good choice.

Determine your soil type. During a recent site assessment, I brought a small shovel and sank it into the ground in locations spread out over

SOIL TYPES

Soil is usually a combination of the following types, like sandy loam or gravelly clay.

Soil type	Description	Meadow plants
Rich	Rich and fertile, tends to be dark brown or black thanks to a surplus of organic matter, like dead and decomposing plants and soil organisms. Rich soil holds moisture well and contains lots of worms and soil critters.	Most plants do well in rich, fertile soil—sometimes too well.
Loamy	This type of soil is less rich and generally grittier. Loamy soils drain easily and include pockets of air, which makes it easier for plants to get the nutrition and water they need.	Most meadow plants like loamy soil, including blue vervain, anise hyssop, rose milkweed, beardtongue, tufted hairgrass, and wild bergamot.
Clay	Clay soil is fairly dense and slippery when wet. When it dries out, it retains its density but becomes chalky and unforgivingly solid. Once this happens, it can be very difficult for clay soil to rehydrate.	Many meadow plants are well suited to clay soil, among them, echinacea, mountain mint, yarrow, meadow blazing star, aster, wild quinine, wild bergamot, and hoary verbena.
Sandy	Sandy soil doesn't retain water well and contains almost no organic matter. It's gritty to the touch and very loose when dry. It can clump some when wet but begins to dry out almost immediately.	Butterfly weed, little bluestem grass, purple love grass, and mountain mint do fine in sandy soil.

the yard. I'd dig five or six inches and observe the soil's color and general characteristics. Was it moist? Did I hit rock? Did it seem loose? Was it heavy or light?

These details all offer up clues about your soil. If your soil is dark-colored and rich, you may want to avoid plants that can become aggressive or grow tall in fertile soil. If it's light-colored, gritty and drains quickly, go for plants that prefer sandy or dry soil. Hard-packed, dense soil tends to have a high clay content, which is difficult for more delicate plants to grow in. Rocky soils are fine for meadow seeds but can be hard to till.

My digging efforts turned up a sandy loam—medium brown soil that mostly held together when I grabbed a handful, and drains well. This meant it wasn't so rich I'd have to worry about plants growing out of control, which can happen in fertile soil. Knowing that this particular soil combination is favored by a wide range of meadow species, I moved ahead on developing a seed mix.

Consider a soil test. If you're concerned about soil acidity, you might want to do a soil test. These tests can determine the pH balance, nutrient content, and level of organic matter in your soil and are available through a state university's cooperative extension office. Follow the instructions on how to collect your samples, send them in, and you'll get results in about two weeks.

If the test finds your soil pH is unusually low, 4.5 or below, for example, you'll want plants that can take more acidic soils. If it's rich in organic matter, it's better to find plants that won't get too rowdy in fertile conditions. A test can fine-tune your plant selection, especially if you're new to meadow-building.

Once you've got a sense of your soil type, weather zone, and rainfall levels you're ready for the next step: designing the right mix of plants for your site.

DESIGN PLANS

Two years ago I had the idea to convert a grassy slope next to Northampton's City Hall into a small urban meadow. The project fell into place quickly. I'd proposed adding the meadow through Local Harmony, the nonprofit I helped start to encourage and practice grassroots stewardship and regeneration, in this case, by creating native pollinator habitats. With the financial support of the city, we bought perennials wholesale and tax-free, and volunteers took care of everything else.

On a single warm November day, our team of fifteen or so meadow enthusiasts, including students, children, and neighbors, cleared the site of weeds and grass and planted more than 750 landscape plugs by hand. Afterward, we celebrated by eating donuts from a local bakery across the street.

After our planting day and well into the following spring, the weather turned unusually wet, and the dry-loving little bluestem grass we'd planted among the flowering meadow plants suffered. It would have managed fine had the roots had a chance to mature. But the young bluestem rotted so badly we were forced to replace almost all of it by late spring. Still, the rest of the plants kept on going. By July, when I dropped by to check up on things and pull some weeds, the echinacea and black-eyed Susans were in full flower.

Then fall arrived and along with it a series of hard-killing frosts. I was unusually busy that year, my schedule packed with work. I'd sold my first design business and the work I'd done building up my new one was paying off. When I finally found time to drive over to the little meadow garden, it was already November.

It had been two months since my last visit and I expected to see a lot of dead husks. Most meadows begin to fade with freezing temperatures and a garden just a few blocks away, with many of the same species, had already turned brownish and wheat-colored. But not this meadow. There, between the street and City Hall, bloomed purple anise hyssop and bright yellow shocks of goldenrod. Watching the late fall sunlight illuminate those flowers, I once again found myself admiring the resilience of meadows, especially when the plants are a good match.

Designing a successful meadow is mostly about choosing grasses and flowers that will be happy where you plant them. This is why doing a site assessment is so important. Meadow plants may be pretty adaptable, but like all plants, animals, and humans, they have strong and particular preferences. If you live in Boulder, Colorado, you'll want plants that can handle cold winter temperatures, low humidity, and clay soil, not plants that grow well along fertile, moist riverbanks in the humid heat of Georgia. Your job as a meadow designer is to play to those preferences. (See "Plant preferences," page 96.)

It takes two or three years for meadow plants to develop the deep roots that will anchor them and help them last. A good design can carry a young meadow through these critical early years, when plants are most vulnerable to disease, drought, and flooding.

Global warming has introduced new challenges, to be sure. The little bluestem didn't make it because of the unexpectedly heavy rains. And the earlier-than-usual arrival of serial cold frosts surprised all of us who'd spent the previous October picking pumpkins under warm, sunny skies.

Fortunately, meadow plants happen to be very robust. Black-eyed Susans have proved, many times over, that they can thrive in dry, hot conditions and put up with hard rain, too—as long as it eventually lets up. And goldenrod, among the hardiest plants, can weather intense heat, bitter cold, and everything in between.

But a meadow's best defense against climate change is the wide variety of species it contains. I've found that a mature meadow can withstand most of what our warming earth is throwing its way. Even if some plants fail, enough other grasses and flowers are around to keep a meadowscape going. In short, a well-designed meadow can take the heat.

THE BASICS OF GOOD DESIGN

A good meadow design takes into consideration your climate, the qualities of the plants you choose, and how tall you want your meadow to be, among other things. You also want plants that will take turns flowering throughout the growing season, so your meadow is always in bloom. One more thing: Trust the process. Plan your meadow, plant it, and then see what turns up. Nature has a way of sorting things out.

Use plants native to your area. The sturdiest plants are ones that have evolved to handle your environment. Natives have also become an integral part of the local ecosystem, like the milkweed that monarch butterflies rely on to complete their life cycles. The ability to support these complex relationships is part of what makes a meadow regenerative.

I generally use plants that grow locally, but I'm not averse to using regional ones because they can be easier to find, which gives me more design options. While I don't plant non-natives because they can become aggressive and overwhelm a meadow, not all of them are bad. Some, like daisies, have naturalized and live in relative harmony with the local ecosystem.

Choose seeds or plugs, or both. The size of your meadow can determine whether to plant seeds or plugs, which influences your design process. I generally sort meadows into two categories: The field by City Hall in Northampton is a *meadow garden*, which means it's small enough to be planted from plugs, which come in trays and are tinier than the average size plants in nurseries. They also grow much more quickly than seeds. Seeds are well-suited to larger *traditional meadows*.

It's easier to design with seeds because they'll do much of the work for you. Seeds are sensitive enough to sort themselves out based on microclimates and small variations in soil quality, moisture, and more. You may find, for instance, that butterfly weed establishes itself in one part of a new meadow because it's slightly sandier there, while the indian paintbrush you planted at the same time sprouts on the other side because it likes the soil density.

Planting from plugs is a more deliberate process, since you're the one who controls where they go. Before ordering any plugs, you'll want to work out a design concept. The goal is to spread the grasses and flowers as evenly as possible, creating a randomized effect while giving each plant enough room to grow to maturity. You can fine-tune the spacing once it's time to plant your plugs.

As for using seeds *and* plugs, I sometimes recommend using plugs in areas you want to grow faster, like a strip along a well-used walkway, and using seeds everywhere else.

Grasses first. Over my many years of planting gardens and meadows, I've come to appreciate that while blossoms come and go, grasses are around for the duration. Grasses give fields the look and feel of a meadow, and are the reason they're often beautiful in the dead of winter. At times I've been so struck by the sight of grass husks iced by frost and glinting in the early morning sunlight I've had to pull over my car to take it in.

Since grasses ground a meadow visually and are the dominant plants, choose them first. Keep in mind that they need to coexist with neighboring meadow flowers; you don't want grasses that spread too aggressively and will force out the other plants.

Grasses tend to be categorized as *cool-season* or *warm-season* plants. Cool-season grasses grow quickly earlier in the growing season when daytime temperatures average 60 to 75 degrees Fahrenheit. In the full heat of summer, they often go dormant, or stop growing, to conserve energy. This type of grass is regularly used for lawns, often includes non-native species, and tends to form into mats. The density

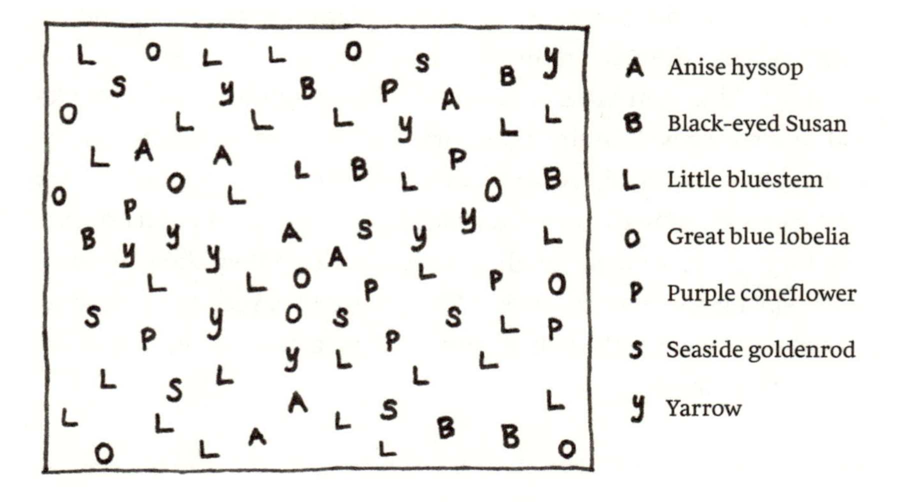

Before ordering your plugs, think about where you want them to go. This is the plan I put together for the Northampton meadow near City Hall. To spare yourself the chore of penciling in hundreds of little plugs, carve out a sample section of your meadow—10 percent makes the math easy. Calculate 10 percent of your grass plugs and space them on your map. Do the same thing with 10 percent of each perennial flower species, and intersperse them with the grasses for a natural look. You can adjust the positioning when it comes time to actually plant your plugs.

of their roots explains why meadows made of cool-season grasses tend to support fewer wildflower species.

If you live in a very cold part of the country, say in Wyoming, Minnesota, or northern New England, you may want to rely more on cool-season grasses. But for most of the country, warm-season grasses work best. These are the most common native meadow grasses and make up the backbone of my meadows. They do most of their growing when temperatures regularly reach above 75 degrees Fahrenheit and usually grow in clumps, which makes room for other plants.

I include at least two, and as many as three or four, warm-season grasses in my meadows if I'm planting a larger meadow. I tend to use one or two perennial grass species, and no more than three in a small one. No matter how large the meadow, I make sure at least 35 percent and as much as 75 percent of the plants are grasses so it looks meadow-like. The exact ratio is really up to you.

Little bluestem grass is my go-to when conditions warrant, like soil that's very well drained and dry. Switchgrass, side oats grama, and prairie dropseed are also excellent warm-season grasses. But cooler-season grasses have their place, too. I often add at least one, like Virginia wild rye or nodding fescue, because they show up early in the spring and mature quickly. This means they keep your meadow looking green when the rest of your young grasses and perennials are just getting started.

Find plants that are about the same height. It's much easier to achieve a lasting balance this way. Most of my meadows are between two and four feet tall. While taller plants have their place, particularly in large spaces or prairie restoration projects, I rarely use them in smaller, suburban meadows because they can overwhelm a house and make a field look messy. Taller plants can also be more prone to flopping over in heavy rains or wind, leaving behind tired-looking, flattened scapes.

I occasionally bend my own rule and include a few shorter flowers if I think they can add a splash of color without getting lost or taken over. Another way to manage a mix is to separate out seeds for taller plants, like perennial sunflowers or Joe-Pye weed, so you can plant them along your border or in a corner and they don't detract from or dominate the rest of your field.

Choose a variety of flowers. I typically include fifteen to twenty-five flowering species when planting a large meadow from seed, and make sure at least one or two will be blooming at all times. If a few of them fail to sprout, no problem. You have enough back-up perennials to take up the slack.

You won't need as many different flower species if you're planting a smaller meadow garden, because it will look too busy and chaotic.

This holds true whether planting from plugs or seeds. In small meadows, I usually include between seven and fifteen flowering perennials.

Pick colors that look good together. Finding flowers to match site conditions is an important consideration, but so is choosing what you like. I know one person who was open to every color but white. She couldn't stand the thought of a field of white flowers, so we didn't plant any. Another client had a just-big-enough-for-a-meadow clearing that barely allowed in the requisite half day of sunlight. He gave me the go-ahead to bring on a riotous blend of bright red bee balm and cardinal flowers, as well as purple coneflowers, perennial sunflowers, and deep orange butterfly weeds. A woman who lives in a purple-gray ranch house wanted to accent it with purple, pink, blue, and white flowers, and it really did improve the overall look of the house.

Sample Meadow Designs

The size of your site greatly influences your design, and the following treatments illustrate just how different the solutions can be. One design is for a meadow garden in western Massachusetts (the species work in most of the country east of the Rockies). It's so small I used

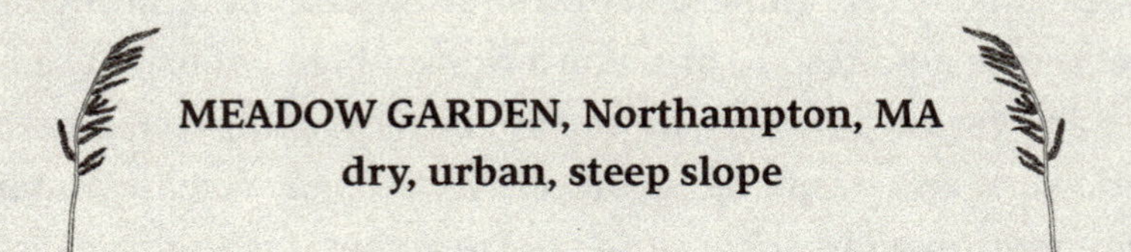

MEADOW GARDEN, Northampton, MA
dry, urban, steep slope

Site description: A generally healthy lawn, not chemically treated, located on a slope so steep the people who mowed it had to wear golf cleats so they wouldn't slip. The planting area is wedge-shaped, about forty-five feet long and twenty-two feet wide on one side, and sits in direct sunlight all day long. The soil is an average-quality loam that drains quickly and becomes very dry by midsummer.

Design solution: I chose little bluestem grass because it can withstand drought. I supplemented it with purple coneflower, anise hyssop, seaside goldenrod, great blue lobelia, black-eyed Susans, and yarrow—six flowering species in all. Each could handle the site conditions and would take turns blooming over the summer. I'd like to have included three or four more flowering perennials but our budget was tight. After planting the plugs, interspersing the grass more or less evenly with the flowers, we mulched. The partially decomposed bark helped control the weeds and protect the soil against erosion, even after many heavy rains. The plants matured quickly and filled out the space within a year.

only plugs and included a handful of species. The other is for a traditional midwestern meadow planted from seeds. With seeds, once you've figured out the size of your lot and the type of soil, plus how tall you want your meadow to be, the rest mostly falls into place.

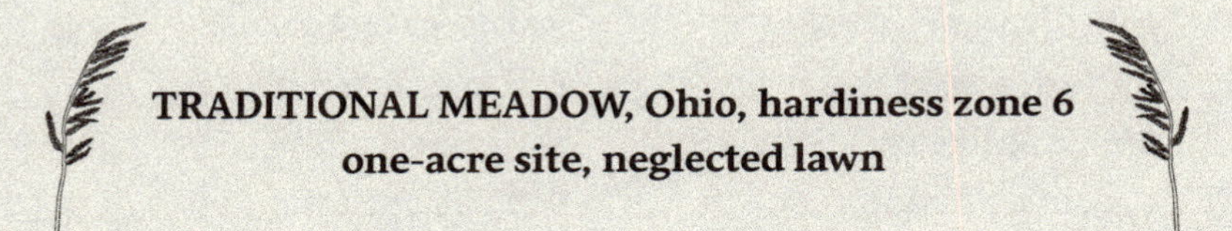

Site description: The entire space was lawn but it hadn't been chemically treated in many years. Since it's a large site, seeds made more sense than plugs. The soil is fairly dense and clay-like, but still somewhat fertile. The area is level but drains well, except for the few weeks a year during the spring thaw. It's fairly exposed and gets full sun and lots of wind.

Design solution: As with all meadows, I started with the grasses because they provide the backdrop for the bright colored flowers. I chose the following grasses because they're two to three feet tall, grow well in regions that get regular rainfall, and can handle dry conditions for long stretches of time. I added the flowering plants for their range of bloom times, ability to thrive in clay soil, and because they provide lots of nectar for pollinators. (Roughly half of the following plants are profiled in more detail in the next chapter.)

GRASSES

Side-oats grama grass	*Bouteloua curtipendula*
Virginia wild rye	*Elymus virginicus*
Little bluestem	*Schizachyrium scoparium*
Prairie dropseed	*Sporobolus heterolepis*

FLOWERS

Yarrow	*Achillea millefolium*
Anise hyssop	*Agastache foeniculum*
Butterfly weed	*Asclepias tuberosa*
Showy milkweed	*Asclepias speciosa*
Cream wild indigo	*Baptisia bracteata*
Indian paintbrush	*Castilleja coccinea*
Lance leaf coreopsis	*Coreopsis lanceolata*
Purple coneflower	*Echinacea purpurea*
Rattlesnake master	*Eryngium yuccifolium*
Cream gentian	*Gentiana flavida*
Stiff gentian	*Gentianella quinquefolia*
Gayfeather	*Liatris spicata*
Great blue lobelia	*Lobelia siphilitica*
Lupine	*Lupinus perennis*
Bee balm	*Monarda didyma*
Wild quinine	*Parthenium integrifolium*
Hairy Beardtongue	*Penstemon hirsutus*
Short tooth mountain mint	*Pycnanthemum muticum*
Showy goldenrod	*Solidago speciosa*
Aromatic aster	*Symphyotrichum oblongifolium*
Hoary vervain	*Verbena stricta*

MEADOW PLANTS

Part of the joy of establishing your own meadow is getting to know your plants. As you do, you'll quickly see that each and every one has its own personality and unique needs, and its own astounding power. When you build a meadow you're actually installing a solar-powered, regenerative system that stores carbon while creating a wild habitat. Even humanity's most brilliant technologies can't come close to pulling this off.

Most of us relate to plants through their look, smell, and feel. But even as a kid I knew that plants, like people and animals, are more than their physical attributes. Certain types of mint are famously gregarious, and I've watched Canada goldenrod happily take over a field like an overenthusiastic house guest. Growing up I observed that the little yellow trout lilies flowering in Maine's still-chilly spring grew in the ditch near our house, and only there. They never moved into the neighboring woods or fields, where they might encounter more plant competition or less moisture, which struck me as very discerning.

Plants are defined in part by the relationships they have with other plants, and with animals, too. Native Americans consider animals to be the original herbalists. I've seen porcupines carefully select young raspberry leaves as if they knew they were loaded with minerals and antioxidants. Once, in my own yard, I watched a young black bear gorge on dozens of apples and then purposefully harvest wild lettuce, a natural pain-reliever, as if warding off a potential belly ache. I've also seen a rabbit dig a shallow hiding spot in the rich soil of a garden bed and carefully line it with thyme, a powerful anti-parasitic that can fend off fleas and ticks.

As with every living thing, plants aren't fixed objects and sometimes their particular properties and tendencies can be hard to pin down. The exact same species can behave quite differently in different locations and when exposed to varying soils, rainfall levels, and light. Switchgrass that grows in very dense soil on a roadside in South Dakota, for example, is going to be shorter and sparser than the same grass growing on a rich, loamy floodplain in Alabama. To the untrained eye, these two plants might not even look like the same species.

In the pages to follow, I introduce you to twenty-one grasses and flowers, all good starter plants for meadow builders everywhere. They're well suited to meadows in yards and near homes because they mature relatively quickly, aren't too tall, and tend to have long bloom times. Each is a perennial, and regenerates through self-seeding or spreading, or both, which helps your meadow store carbon more efficiently while also making it long-lasting and easy to maintain.

A few things to keep in mind as you read through them: All the grasses are warm-season ones, the most common native grasses in meadows. I've listed each plant's 'hardiness zone' to help you choose the right varieties for your climate. I've also indicated the 'bloom time,' the window of time during which a plant can flower, depending on where you live. (It does not suggest how long a plant flowers!) As mentioned, the personality of each plant depends somewhat on where it's grown.

These species represent a fraction of the hundreds of others thriving in meadows across the country. Consider using them to start the process of creating your own meadow blend, then go ahead and explore—filling in the gaps with whatever local grasses and flowers most inspire you. I hope you enjoy getting to know these very cool plants as much as I have.

Plant profiles

GRASSES

FLOWERS

BLUE GRAMA GRASS

Bouteloua gracilis
Height: 1 to 1½ feet
Hardiness zones: 3 to 10
Bloom time: July, August, September

This rugged plant was a staple of the shortgrass prairie that covered the Great Plains, and along with buffalo grass, was grazed on by herds of bison well before white settlers arrived. Blue grama grass is the taller, more decorative of the two. Nicknamed "mosquito grass," it has delicate, bluish-green seed heads that resemble misplaced false eyelashes and seem to hover above its stalks like a cloud of insects.

Blue grama grass is very drought-tolerant thanks to its long, vigorous roots. That and its resilience made it an important part of this country's recovery effort after the Dust Bowl, when it was used to restore exposed soils. It's not a good choice for areas that get lots of rainfall because it will be easily overtaken by faster-growing, moisture-loving meadow plants. But in dry locations where it can handle the competition, this grass grows like a champ. It's very reliable and requires virtually no maintenance.

I use this grass in small meadow gardens as well as in bigger meadows grown from seed. It can be paired nicely with other low-growing grasses and flowers, such as little bluestem, butterfly weed, yarrow, and coreopsis. Another reason to like blue grama grass: It injects color into pale winter landscapes. The semi-evergreen grass lightens over the summer and is tawny-colored by fall.

WHERE IT THRIVES

Resilience. Blue grama grass is able to survive very dry conditions better than most grasses thanks to its robust root system. It can also put up with extreme cold.

Regional compatibility. This short grass is native to dry zones that extend from southern Canada through Texas and over to southern California, but it can grow in most places that tend to be dry. It doesn't do as well in humid environments, like the mid-Atlantic region or the Southeast.

Soil conditions and sun. Blue grama tolerates many soil types, including clay, sand, or gravel. It can handle low-quality or average soil, but is happiest in soil that's well drained. It does not need rich soil to thrive, but tends to grow more densely in higher-quality, drier soils. This perennial loves sun but can handle shade—it just won't be as full.

PLANTING AND GROWING

Plant blue grama in the spring and you can expect to see seed heads emerge by midsummer. This grass spreads slowly from rhizomes, or underground roots. It's fairly pest-resistant and can survive being grazed by deer or other animals. But it doesn't like soggy or humid conditions, which make it vulnerable to rust, fungal rot, and smuts.

PURPLE NEEDLE GRASS

Nassella pulchra (formerly Stipa pulchra)
Height: 1 to 3 feet
Hardiness zones: 5 to 10
Bloom time: June, July

Purple needle grass is a classic California meadow plant, complete with purple-tinged blades that sway in spring breezes and turn a golden wheat color come summer. I'm highlighting it for western readers because this hardy tufted grass, also known as purple tussock grass, is really useful. It's often used in land restoration projects because it's long-lived, grows even in poor soil, and needs very little water. It can also help prevent erosion thanks to a strong root system, which grows as deep as twenty feet in search of moisture.

Considered California's state grass, purple needlegrass used to cover large swaths of the state. Sadly, that's no longer true thanks to competition from non-native grasses, which now dominate local grasslands and agriculture. Planting this grass can give local ecosystems a boost and attract native California butterflies and moths, including the Juba skipper, common ringlet, Nevada skipper, and Uncas skipper, as well as many varieties of birds.

WHERE IT THRIVES

Resilience. This perennial's deep root system makes it very resilient and tolerant of extreme heat and drought, which is why it's well suited to California.

Regional compatibility. *Nassella pulchra* is unique to California, but its cousins, which do better in moist conditions and can tolerate occasional flooding, can be grown elsewhere. Green needle grass (*Hesperostipa viridula,* formerly *Stipa viridula*), for example, is native to the Rockies. Porcupine needle grass (*Hesperostipa spartea*, formerly *Stipa spartea*) grows naturally on prairies in the central part of the US, and can be planted in the eastern part of the country, too, especially in colder regions.

Soil conditions and sun. It prefers dry conditions but needs spring rains to thrive. It doesn't do well in soils that are moist for long stretches of time. Purple needle grass likes full sun and won't do well in shade.

PLANTING AND GROWING

Be sure to seed purple needle grass into a cleared site. Because it grows slowly and isn't very tall, it can otherwise be quickly overtaken by weeds and exotic grasses. It needs rain, but not a lot, to germinate and take root. Once established, it reseeds itself prolifically and will stick around for a long time. Grazing animals will eat the grass when it's young and tender. It can handle some grazing, but if it becomes a problem you can use natural pest control methods to keep animals away. (See "Organic pest controls," page 125.)

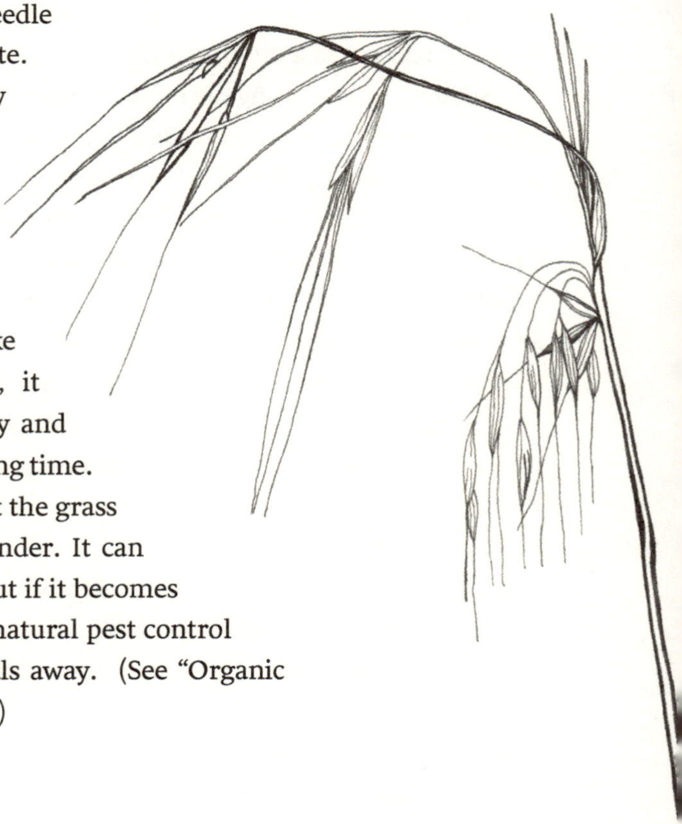

SWITCHGRASS

Panicum virgatum
Height: 3 to 6 feet
Hardiness zones: 3 to 9
Bloom time: August, September, October

I use switchgrass in many of my meadows because of its reliability and good looks. It produces tiny pink-tinged flower spikes that hover above the green stalks. Ripe seeds may be pink or dull purple at first, but turn a golden brown in autumn. The grass usually stays upright throughout the year, lending beauty to an otherwise muted winter meadow.

Switchgrass is among the most common species of native tallgrass prairie grasses, and is also called tall panic grass, tall prairie grass, wild redtop, or thatch grass. It's a warm-season perennial, meaning most of its growth takes place from early summer through early fall. In northern cold regions the productive season can be as short as three months, but in southern areas, like along the Gulf Coast, its growing season can last eight months and it's usually taller.

Switchgrass's roots can reach several feet deep. This robust root system makes it very helpful in controlling erosion and supporting essential soil microorganisms, and sinking lots of carbon.

WHERE IT THRIVES

Resilience. Switchgrass is highly adaptive. It can be found growing on sand dunes and in mine tailings, as well as in healthy prairies. Once established it can tolerate some drought and grows well in high temperatures.

Regional compatibility. Thrives in most places except arid zones, since it needs moisture year-round.

Soil conditions and sun. Switchgrass is well suited to clay and sand-heavy environments but will suffer without much water. In rich, moist soil it can grow tall and fill out substantially. Switchgrass requires full sun and becomes sparse if it gets too much shade.

PLANTING AND GROWING

This plant germinates quickly from seed but takes about three years to mature. You can plant faster-growing plugs any time during the growing season. Once mature, switchgrass reaches its full height early in the summer and seed heads appear in late summer.

You can check switchgrass growth by planting it with other vigorous meadow plants of similar height. If it becomes too outgoing, you can limit its spread by selectively mowing it in the summer using a scythe or weed wacker. Deer and other wild ruminants might also help trim it, which is fine; as with most grasses, switchgrass can handle some grazing.

TUFTED HAIR GRASS

Deschampsia cespitosa
Height: 2 to 3 feet
Hardiness zones: 3 to 8
Bloom time: June, July, August, September

Whenever I introduce this grass to a client, they're invariably blown away. Grass may not seem exciting, but then you haven't seen the way low sunlight lights up the delicate billow made up of this plant's tiny seed heads. All those florets and seed heads make the grass glisten in dew and frost. Even when not producing dazzling effects, the grass's tawny stalks and seed heads spruce up a meadow all winter long.

This species of grass grows almost everywhere—from tidal marshes to mountain meadows. It's also a grass that gives back more than most. It can revegetate rocky alpine zones, feed wildlife at high elevations, and reclaim mining sites, thanks to a resistance to toxins. It even feeds bears. Tufted hair grass is among the few perennial grasses that prefers shade.

WHERE IT THRIVES

Resilience. This grass doesn't do well in high heat or lots of sun, but in cool conditions, it's a very tough plant and self-seeder.

Regional compatibility. Tufted hair grass is native to the western United States as well as the northern plains and the Northeast. In the wild, it's most often found in moist to very wet areas, such as freshwater or saltwater marshes and woodland habitats. *Deschampsia flexulosa*, or wavy hair grass, is an eastern cousin that behaves similarly.

Optimal soil conditions and sun. This grass likes average to moist conditions. It prefers some shade, but can tolerate full sun if mixed in with other meadow plants that can protect it.

PLANTING AND GROWING

This plant is fairly easy to establish from seed. It tends to flower more abundantly in northern and high-elevation meadows, and may not bloom in the Southeast and other warmer regions. Tufted hair grass has virtually no pest or disease problems. As with most meadow plants, it benefits from being cut back in early spring.

ANISE HYSSOP

Agastache foeniculum
Height: 2 to 4 feet
Hardiness zones: 4 to 8
Bloom time: June, July, August, September

Pollinators adore this plant, which is one reason I'm drawn to it. Last fall, I was finishing a project near some anise hyssop that still had a few blooms. It's a time of year when nectar's in short supply, and I got to watch a scattering of late-season butterflies sipping from the light purple flowers.

I use this perennial herb a lot. Along with its lovely flowers, it's reliable and durable. It's also edible. The dried leaves can be brewed into a soothing tea, and the fresh leaves used to flavor jellies or salads. A member of the mint family, anise hyssop is used in Western herbal and traditional Native American medicine to treat coughs, fevers, wounds, and diarrhea, and I suspect its antiviral and antimicrobial properties may also benefit the health of pollinators and their hives.

Anise hyssop's minute, pale purple flowers cluster in dense spikes. It can bloom anytime from early summer all the way to the first frost. Once it begins flowering, it usually blooms for around two months.

WHERE IT THRIVES

Resilience. This highly adaptable plant does well in the most inhospitable soils and weather. Once established, it tolerates drought well.

Regional compatibility. The flower is native to the northern parts of the Midwest. It is hardy in the coldest regions, but can also thrive in warmer regions up to and including zone 8.

Soil conditions and sun. Anise hyssop thrives in dry soil, but can tolerate moist soil as long as it's well drained. In really rich soil it can get taller, reaching four feet, or even higher. It favors full sun but can handle some shade.

PLANTING AND GROWING

This plant tends to bloom quickly, often in its first full season of growth. It spreads by growing underground horizontal roots and easily self-seeds. If you have a small meadow garden and want to keep it in check, pull any seedlings you don't want. To prevent prolific self-seeding and promote heavier blooming, you can also pinch, or deadhead, the spent flowers. This plant may develop root rot in wet soils or powdery mildew and leaf spots in humid climates. It's fairly pest-resistant. Deer find the smell and taste of the plant repulsive.

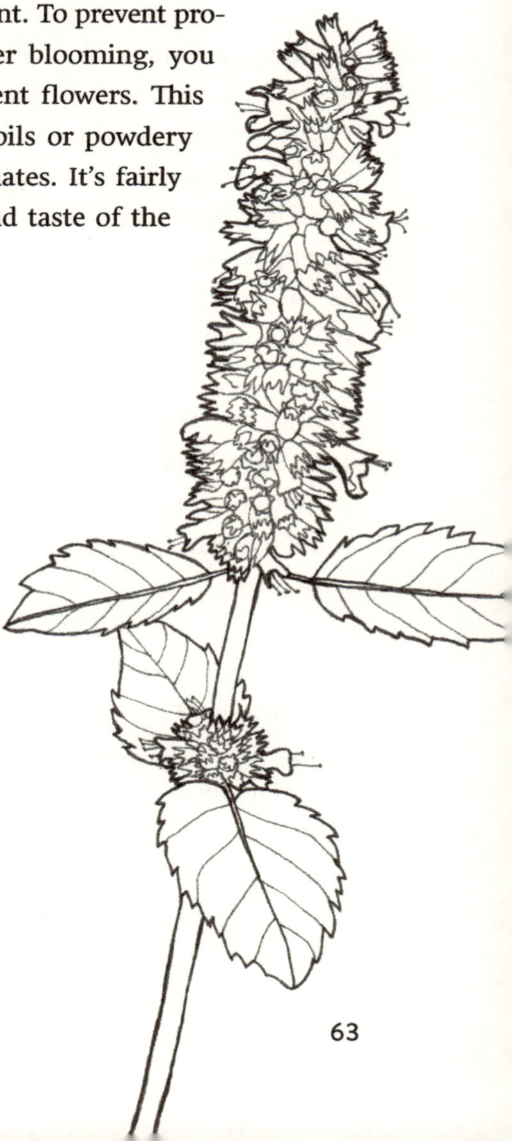

BLACK-EYED SUSAN

Rudbeckia hirta
Height: 1 to 3 feet
Hardiness zones: 3 to 9
Bloom time: June, July, August, September, October

Black-eyed Susan's distinctive, cheerful flowers will be familiar to even the most inexperienced gardeners. I can remember seeing them in bouquets my mom made when I was a kid, and watching them bloom in the field across from my house where we picked wild strawberries in the summer and apples in the fall.

This bright perennial is known as a "pioneer species," meaning it's often among the first plants to re-colonize barren environments marred by clear cuts, fire, or other ecological disruptions. It's also easy to grow, which makes it easy to recommend to first-time meadow makers. These long-lasting flowers are a great addition to any meadow design because they perform predictably in all the best ways.

WHERE IT THRIVES

Resilience. This super-tough plant can acclimatize to heat, drought, and poor soil conditions. It doesn't like to have its feet wet too often, so in wetter locations, it can rot without plenty of sun and good drainage.

Regional compatibility. Black-eyed Susans are native from the Rockies eastward, and their bright yellow color is a common sight in meadows across the country.

Soil conditions and sun. They prefer acidic soils but can thrive in just about any soil, as long as it's well-drained. These flowers do fine in partial sun, they just won't fill out as much or flower as readily.

PLANTING AND GROWING

Black-eyed Susan seeds tend to germinate quickly. They sometimes flower during their first full season of growth, and can bloom twice in longer growing seasons. One caveat: If given too much space and no competition, black-eyed Susans can become aggressive and crowd out neighboring flowers, especially short ones. If you plant them in a smaller meadow or meadow garden, keep an eye on them to make sure they don't take over.

BUTTERFLY WEED

Asclepias tuberosa
Height: 1½ to 2 feet
Hardiness zones: 3 to 9
Bloom time: June, July, August

This colorful species of milkweed is known for its ability to attract monarch butterflies, as well as other types of butterflies, bees, and hummingbirds. Though sometimes called orange milkweed, it has no milky sap. Long-lived and hardy, this beautiful plant grows clusters of tiny flowers that bloom in stunning patches of bright orange between June and August. By the fall it develops long pods containing hundreds of seeds with tufts of long, silky hairs, which I love seeing in wintry fields.

I recommend butterfly weed for meadows located in drier or very well-drained conditions. It has even done well for me in hot, urban environments.

WHERE IT THRIVES

Resilience. Once established, butterfly weed tolerates drought and hot weather.

Regional compatibility. This plant is native to the Southwest and all regions east of the Rockies. It grows just about everywhere in the United States.

Soil conditions and sun. This dry-loving plant can take some moisture as long as the soil is very well drained; it won't grow if its roots are often wet. While it prefers sandy soil, it also grows well in clay soil and dry or rocky soil. It won't flower much if it gets too much shade.

PLANTING AND GROWING

Butterfly weed seeds germinate quickly and grow slowly. It can take two or three years for them to flower. If planting from plugs, be sure not to overwater them. This plant usually forms clumps, which means you might see it growing in distinct patches throughout your meadow. It has a long taproot that can be easily damaged if you try to transplant it, so don't.

Patches of butterfly weed may be attacked by aphids, but as long as there aren't too many, your plant should be fine. To get rid of them, apply an insecticidal soap or simply hose them off. Deer are not attracted to this plant.

EARLY SUNFLOWER

Heliopsis helianthoides
Height: 4 to 6 feet
Hardiness zones: 3 to 8
Bloom time: June, July, August, September

This undemanding plant resembles its namesake, with cheerful yellow petals evenly distributed around a round, yellow, helical center. But the early sunflower is actually a member of the aster family. Unlike the common sunflower, its seeds are inedible for humans, but birds love them. It can tower over many other meadow plants and rarely flops over. Its height makes it a better choice for large meadows than small ones.

The early sunflower is extremely resilient and you can find it blooming in wetlands, on dry hillsides, and along freeways. It grows well in dry locations, in poor to average soils, and in full sun to partial shade. Ground nesting bees love this plant, and so do butterflies and other pollinators.

WHERE IT THRIVES

Resilience. Early sunflower prospers where more delicate plants struggle. It can acclimatize to most hardships, with the exception of saline soils.

Regional compatibility. Its native range is east of the Rocky Mountains, but it can grow well in the western part of the country, as long as it gets some rain during the growing season.

Soil conditions and sun. This sunflower likes dry and arid soil, but can thrive in nutrient-poor soil types and in medium-wet soil, too. It grows taller in high-quality soil. This plant prefers full sun, but does fine in partial sun.

PLANTING AND GROWING

Early sunflower seeds tend to germinate quickly, usually within a month if you plant them in the spring. Plant these seeds in the fall and they'll sometimes even sprout several weeks after the first heavy frost. Flooding or high humidity can limit the growth of these hardy plants, but otherwise early sunflowers will almost certainly thrive, sometimes too well! They reseed themselves each year, so in a smaller meadow garden be sure to pull any unwanted seedlings. The plants don't have any real pest problems, and deer tend to steer clear.

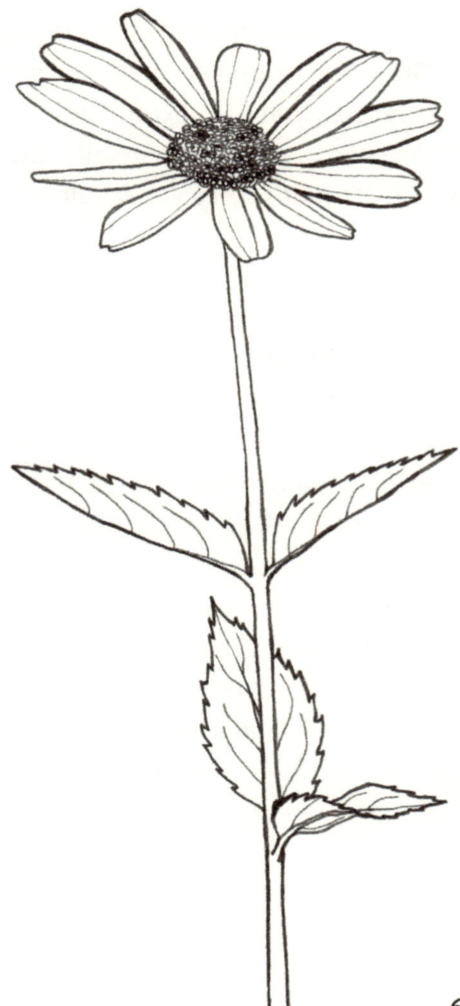

FOXGLOVE BEARDTONGUE

Penstemon digitalis
Height: 2 to 4 feet
Hardiness zones: 3 to 8
Bloom time: May, June, July

Foxglove beardtongue almost always finds its way into my meadows. The colors usually work with whatever else is blooming, and in general, this plant stays in its own lane. Its beautiful, long-lasting, soft pink and white blooms are like a snapdragon's, but more low-key.

The flower blooms early and leaves behind seed capsules that start off a soft red, and add texture and color to a meadow well into the fall. Although it shares the name of the notoriously toxic common foxglove (*Digitalis purpurea*), foxglove beardtongue is not poisonous.

WHERE IT THRIVES

Resilience. Foxglove beardtongue is less drought-tolerant than many other wildflowers but does okay in dry conditions, for short spells. Highly adaptable, it regularly thrives in difficult spots, like alongside busy roadways and railway tracks.

Regional compatibility. Native to eastern parts of the country, it also grows well in the rainy Pacific Northwest and any region that gets regular rainfall. Showy beardtongue and hairy beardtongue are among its many cousins, that do well in meadows.

Soil conditions and sun. Prefers well-drained soil but is amenable to a variety of soil conditions, including clay or richer soils. It can also handle sandy soils in areas that get plenty of rain. Unlike most meadow plants, this perennial can actually do pretty well in partial shade.

PLANTING AND GROWING

Foxglove beardtongue flowers more abundantly with every passing year. Buds appear in the late spring and bloom until the full heat of the summer kicks in. This species actively self-seeds, so in smaller meadow gardens you may want to thin out new sprouts as they come up. Foxglove beardtongue is fairly resistant to pests and ignored by deer and rabbits.

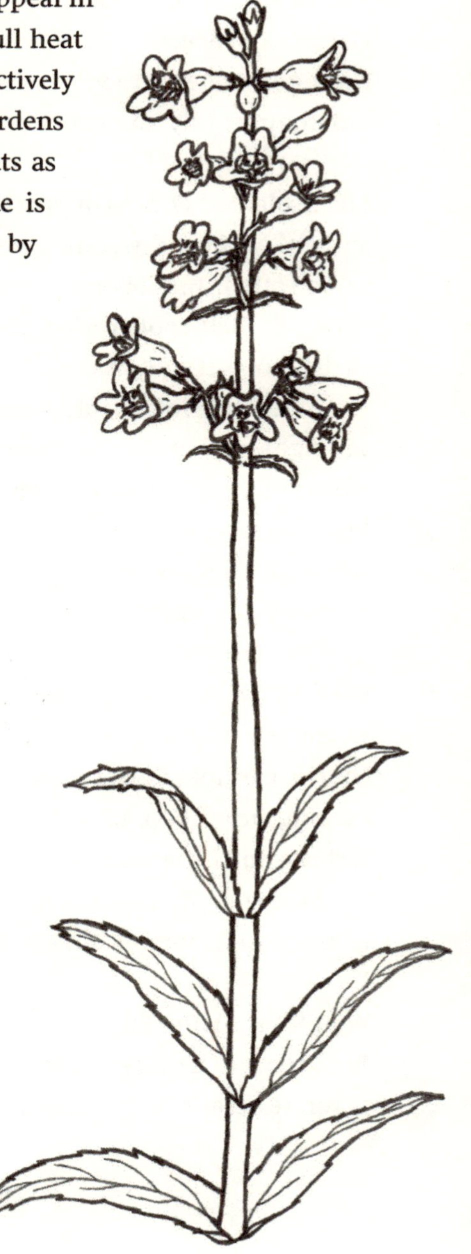

GREAT BLUE LOBELIA

Lobelia siphilitica
Height: 2 to 3 feet
Hardiness zones: 4 to 9
Bloom time: July, August, September

I once planted dozens of great blue lobelia plants in a moist meadow garden. Many did well but when it rained, the plants in the wettest areas rotted. I stubbornly replaced them with a second round. No surprise, their roots rotted, too. So take it from me—this plant likes moist, not wet, conditions.

Great blue lobelia plays nicely with other meadow plants. Give it the right kind of soil and it will self-seed prolifically, growing everywhere in your meadow. If your meadow is small and your great blue lobelia becomes a little too happy and widespread, you may want to weed out a few seedlings once in a while. I tend to let them be because I really love seeing these vibrant, pure-blue flowers.

The plant's tall, tubular flowers, which first appear in midsummer, are usually dark blue, light blue, mixed, and occasionally even white. If you're looking for a red flower, consider great blue lobelia's cousin, cardinal flower *(Lobelia cardinalis)*. Both flowers give a meadow significant pop, plus they continue to supply nectar to bees, butterflies, and hummingbirds when other flowers have faded.

Fun fact: This lobelia gets its botanical species name, *siphilitica*, from the once-upon-a-time belief that it was a cure for syphilis. It's not.

WHERE IT THRIVES

Resilience. This flower does well in moist conditions and can also tolerate occasional periods of drought. But don't plant it in arid parts of the country or anywhere else it won't get watered fairly regularly. It won't be happy.

Regional compatibility. Great blue lobelia grows well in areas that are moist but not overly wet. Its natural range is in the eastern part of the country but it'll grow anywhere that gets regular rainfall.

Soil conditions and sun. Tolerates medium to moist soils, which is why it grows well along river banks, ponds, and streams. But it does fine in chalk, clay, and sandy soil, too. This plant prefers sun but does fine in partial shade. If part of your meadow doesn't get a lot of sun or includes trees, this plant could be for you.

PLANTING AND GROWING

Lobelia is another easy-to-grow plant. If you plant it in a meadow garden and find that it gets too bushy for you, divide the clumps in the spring. The plant is fairly disease-resistant, but not pest-resistant: Moisture-loving snails and slugs love it. There's no reason to fear rabbits and deer, though. They don't like the taste of this plant.

HOARY VERBENA

Verbena stricta
Height: 2 to 3 feet
Hardiness zones: 3 to 8
Bloom time: July, August, September

This tall, spiky wildflower often grows in places where few other plants want to live, like overgrazed meadows, highway meridians, and trash-strewn lots. It's so unfussy and resourceful it's seen as a weed by many, especially farmers. But verbena is beloved by pollinators, and looks beautiful in a meadow with other plants around it. Its toughness also makes it a valuable meadow builder.

In traditional meadows, I plant it with flowers and grasses of similar height. In smaller gardens, you need to be prepared to weed out excess. I once planted it in a medicinal demonstration garden, where it was exceedingly happy—so much so that after its first season, I spent a couple of hours pulling out thousands of tiny seedlings.

Hoary verbena's floral spikes bloom from the bottom up, which means the flowers last over a period of weeks. Even before any purple (and sometimes white) flowers appear, verbena's fuzzy leaves host butterfly larvae throughout the growing season.

WHERE IT THRIVES

Resilience. Hoary verbena is very drought-hardy and can adapt to otherwise hostile conditions, which is why you'll find it growing along roadways and in abandoned lots. If your soil quality seems very poor, chances are verbena will do fine there.

Regional compatibility. This plant is native to most of the US, except for California, Oregon, and a few Atlantic coastal states. It has naturalized, and readily adapted to ecosystems everywhere without becoming invasive.

Soil conditions and sun. The ideal soil for verbena is dry, sandy, or gravelly, but it can handle most well-drained soils. In rich, loamy soil it runs the risk of getting crowded out by neighboring plants. This plant prefers full sun but can tolerate partial shade.

PLANTING AND GROWING

Hoary verbena is easy to grow from seed and reproduces primarily by reseeding itself. The root system consists of a taproot, and in the spring new growth emerges from the top. This plant needs sun and a little rain, and that's it. It is virtually pest- and disease-free.

LANCELEAF COREOPSIS

Coreopsis lanceolata
Height: 1 to 2 feet
Hardiness zones: 4 to 9
Bloom time: May, June, July

Lanceleaf coreopsis is a great example of the scrappy nature of meadow plants. Although it doesn't grow very tall, it can be prolific and quickly spread its bright, daisy-like flowers across a field even in less-than-ideal conditions. I often plant it with other low-growing meadow plants that prefer similarly dry conditions, like butterfly weed and little bluestem grass. Just as humans befriend people who live nearby and share the same interests, certain cliques of plants simply like to hang out together.

I've always been intrigued that this species of coreopsis, as well as its other native cousins, is rather fragile-looking. Its leaves and stems are thin, and the flower petals no thicker than a thin piece of paper. This delicate appearance belies its tenacious behavior, and always seems like a trick the species somehow pulls off.

In autumn, the blossoms fall away to reveal pods loaded with tiny, tick-like seeds, which explains the plant's nickname, tickseed.

WHERE IT THRIVES

Resilience. Lanceleaf coreopsis is famously hardy and can stick it out through periods of heat, drought, and humidity. Not only can it tolerate these conditions, it likes them.

Regional compatibility. Native to the Southwest and Midwest, lanceleaf coreopsis has naturalized everywhere, which means it can be planted just about anywhere in the US.

Soil conditions and sun. This is a good plant for sites with poor, dry soils. It does decently in richer, moist soils, too, but cannot tolerate wet conditions or too much moisture for very long. Lanceleaf coreopsis is unhappy when it receives anything less than a half day of direct sunlight.

PLANTING AND GROWING

This plant grows quickly from seed and plugs. If planted in the spring, it attracts birds, bees, and butterflies by summer, but blooms more robustly in its second year. The flower has few problems with insects or disease. Its only real challenge is that it can be overshadowed and crowded out by taller species. But given the right conditions and room to grow, these plants can easily self-propagate and even take over a field. Lanceleaf coreopsis can become gangly and sprawl if grown in soil that's too moist or fertile.

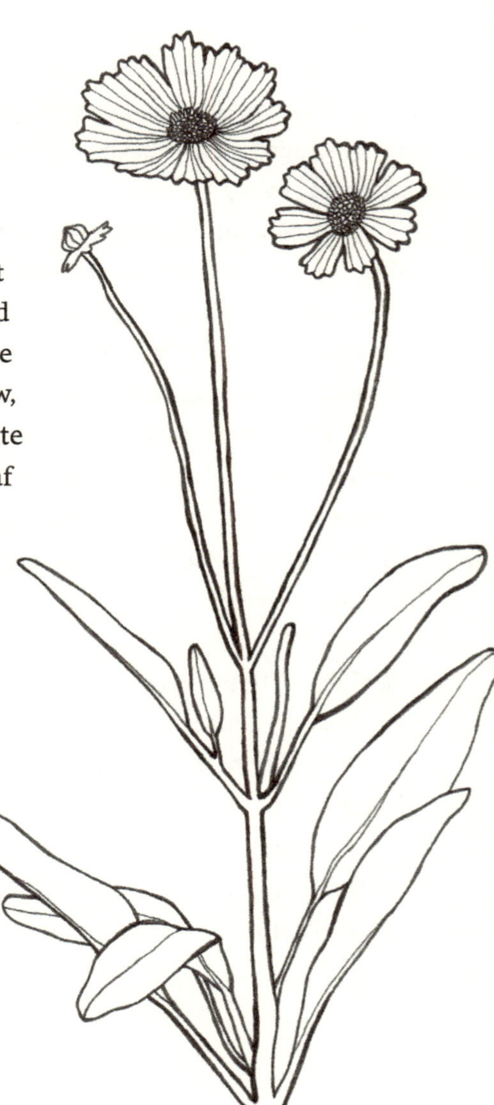

LUPINE

Lupinus perennis
Height: 1 to 2 feet
Hardiness zones: 3 to 8
Bloom time: May, June

Whenever I think of lupine, my mind wanders to the north side of the White Mountains in New Hampshire, a location where these wildflowers thrive. In the town of Sugar Hill, which hosts a popular lupine festival, acres of this perennial reliably turn old hayfields into carpets of purple every June.

Lupine's spires of pea-like flowers are mostly blue or purple, but they can also be white or pink. Its good looks, plus the ability to thrive in cool, moist locations, makes it a popular choice among eastern meadow builders. The rare Karner blue butterfly, which lives in dry, sandy areas in the East, feeds solely on this plant.

Lupine derives its name from the Latin word for wolf. Early colonists thought this plant preyed or "wolfed" on the soil, but the opposite is true. Lupine is a member of the pea family. Like all legumes, it enriches the soil it grows in by adding nitrogen, which allows it to thrive in poor-quality soil.

WHERE IT THRIVES

Resilience. This much-loved legume is drought-tolerant but doesn't do well in high heat.

Regional compatibility. Lupine are at their best in areas with cool summer nights. *Lupinus perennis* is primarily native to the East Coast, but grows further inland as well. It has many cousins, like the California native arroyo lupine, or *Lupinus succulentus*, which thrives in drier conditions. More than one hundred lupine species can be found west of the Rockies.

Soil conditions and sun. Tolerates sandy, dry soil and can handle short periods of wetness. But it thrives in and greatly prefers soil that is well drained. Lupine likes full sun but can put up with light shade. It won't flower if it gets too much shade —like most meadow plants.

PLANTING AND GROWING

Lupines produce dozens of pea-like seeds per plant, which are dispersed when the pods pop open in late July or early August. By the end of summer, its leaves can look a little worn or even go dormant, which is why I tend to avoid using them in smaller meadow gardens.

As with most meadow perennials, once established this plant requires little attention. Since it can spread quickly and become invasive in situations where it's not a native, use a lupine species matched to your region.

MEADOW BLAZING STAR

Liatris ligulistylis
Height: 2 to 3+ feet
Hardiness zones: 3 to 8
Bloom time: August, September

If you want to see more monarch butterflies this summer, plant some meadow blazing star. The monarch's late summer migration coincides with this flower's peak bloom. It also attracts black swallowtails, painted ladies, and sulfurs, as well as some other really cool-looking insects, like shiny-metallic sweat bees.

I use this plant regularly because it grows easily in a variety of conditions. The flower's small, frilly flowers open from the bottom of a long flower spike, like a slow-motion flame burning up a matchstick. This long, slow bloom makes nectar available for a long stretch of time, which is one reason it attracts so many beneficial insects.

Meadow blazing star, also known as gayfeather and colic root, was once used by Native Americans to relieve indigestion. Its native cousins like northern blazing star and bottlebrush blazing star are all great choices for meadows and meadow gardens as well.

WHERE IT THRIVES

Resilience. Meadow blazing star can put up with summer heat and humidity, and mature plants can tolerate drought conditions.

Regional compatibility. Native to the northern parts of the Midwest, it also grows in eastern and southern regions.

Soil conditions and sun. This flower prefers moist, well-drained soil. In especially rich soil, it can grow up to five feet tall, which makes it prone to flopping over. It can handle poor, dry soil once its root system is well established, and won't grow as tall or flop over as easily. This is a plant that prefers full sun. In shade it won't bloom much, if at all, and becomes much more susceptible to disease.

PLANTING AND GROWING

If you use plugs, it's best to plant in early spring or the fall. Plants grown from seed typically start blooming in their second year. This plant will self-seed if it's happy, which means you may want to weed out extra seedlings in smaller meadows. Meadow blazing star also spreads slowly through corms, or tuberous roots. Over the course of many years, a single plant can become three to five feet wide.

The plant grows more quickly and blooms more vigorously in moist but well-drained soil. Wet soil can lead to root rot, which kills the plant. Too much moisture can also bring on leaf spot or rust, which spot the leaves.

MOUNTAIN MINT

Pycnanthemum virginianum
Height: 2 to 3 feet
Hardiness zones: 3 to 7
Bloom time: June, July, August, September

Though I think almost all native flowers are beautiful, I really appreciate the unique qualities of mountain mint flowers. Little white blooms unfurl from a cluster of flower buds and then bloom for a really long time. Each flower is a magnet for a wide range of pollinators, including native bees, beetles, potter wasps, and pearl crescent butterflies.

Mountain mint is neither a true mint nor a mountain native. It's happiest in damp soil, which makes it a good choice for moist meadows and shoreline plantings. But it's also very adaptable and can be found along gravelly roadsides, and in open woods. This plant can be fairly aggressive, so watch that it doesn't take over your small meadow garden. In larger meadows its spreading ability is typically balanced by other meadow plants, so this isn't as much of a concern.

The flower buds and leaves are edible raw and cooked, and have a mint-like flavor that makes a good seasoning or tea. The dried plant is said to be a good natural insecticide.

WHERE IT THRIVES

Resilience. Mountain mint can tolerate floods and humidity, but it doesn't do well in drought.

Regional compatibility. *Pycnanthemum virginianum* is native across the Midwest and eastern United States, including Virginia, of course.

Soil conditions and sun. Mountain mint is totally fine with almost any type of soil, including loam, clay, gravel, or sand. But it grows most happily when it gets plenty of water. It tolerates shade better than most meadow plants.

PLANTING AND GROWING

Plugs planted in early spring usually flower within their first season. The plant can spread very quickly and easily overtake shorter plants. Even in drier situations where it's not aggressive, mountain mint tends to self-seed into a thicket. If it begins to overcrowd the other plants in your meadow garden, go ahead and thin it by pulling it out by its roots.

Most of the problems that can affect this plant stem from too little water, so plant it where it'll get moisture. When stressed or too dry, mountain mint can suffer from rusts and mildews. Thanks to its strong flavor, this perennial is deer-proof.

OBEDIENT PLANT

Physostegia virginiana
Height: 2 to 4 feet
Hardiness zones: 3 to 9
Bloom time: July, August, September, October

I've long admired obedient plant for its feral qualities. It's a commonly used garden perennial but often seems happiest when left alone, in free-range zones like meadows. This highly adaptable plant produces a flush of pink flowers in late summer that are candy for the eyes, not to mention food for the many pollinators that swarm them.

The obedient plant gets its name from its flower stalk, because when bent it stays there. The snapdragon-like appearance of its flowers also explains one of its other names, false dragonhead. But the plant is not always so obedient. Its rhizomatous root system spreads quickly, and in moist, fertile soil it can quickly overtake a field.

WHERE IT THRIVES

Resilience. The plant adapts well to its environment. It grows along riverbanks and in wet meadows but can also put up with somewhat arid conditions.

Regional compatibility. Native to the United States from the Rockies eastward.

Soil conditions and sun. Prefers moist, slightly acidic soil but also does well in average to poor soil, which can contain its spreading. This plant loves full sun, but tolerates some shade as long as it has adequate moisture.

PLANTING AND GROWING

The seeds tend to germinate quickly and can sometimes bloom in their first season. Its only real requirement is that it get water on a regular basis. Plants spread via rhizomes underground, which you can curb, if you want, by pulling out the shoots. In meadow gardens, be on the lookout for aphids and spider mites, and hose them off if you find them.

PURPLE CONEFLOWER

Echinacea purpurea
Height: 2 to 4 feet
Hardiness zones: 3 to 9
Bloom time: June, July, August, September

It's rare that I design a meadow or meadow garden without using at least a little purple coneflower, commonly called echinacea. There's just something about it I find calming. It is sturdy and graceful at the same time, which speaks to its character. A workhorse of a plant, it also handles whatever nature throws its way.

Purple coneflower can grow in just about any climate and handle even barren soil. The light pink blossoms last for a long time in the full heat of summer, attracting crowds of butterflies and other pollinators, and later, birds. It usually grows to about three feet, and generally gets along well with other plants of similar heights.

The coneflower's daisy-like flower is long-blooming, and I like to mix it with anise hyssop and meadow blazing star because the colors blend well. Herbal healers use the flowers of *Echinacea* species for their antibiotic properties.

WHERE IT THRIVES

Resilience. It has a fibrous root system that grows from a thick rhizome and can extend deep into the soil, helping it withstand drought, heat, and humidity. Once established this plant is impressively resilient.

Regional compatibility. The flower's native range is the eastern half of the United States. It has a number of cousins with a similar range that are also good candidates for meadows.

Soil conditions and sun. Don't worry about soil quality. Average to dry, well-drained, and even rocky soil will do. Just avoid overly soggy or wet conditions for this plant. Purple coneflower prefers full sun exposure. You can grow it in filtered shade, but expect fewer flowers.

PLANTING AND GROWING

Plant this flower from seed when the weather is cool, in the spring or fall. For plugs, the spring is best so they get a chance to bloom in their first season. The summer is fine with a little watering, but not too much. These plants can adapt to just about anything except too much water.

Coneflower is clump-forming and self-seeds easily, which means you'll see patches of it moving around in your meadow over the course of many growing seasons.

It has few pest and disease issues but can become food for slugs when conditions are wet. Deer and other grazing animals will eat young coneflower plants but typically avoid mature ones, unless they're really hungry.

SHOWY GOLDENROD

Solidago speciosa
Height: 3 to 5 feet
Hardiness zones: 3 to 8
Bloom time: August, September, October

I've been fond of goldenrod since childhood because it was one of the first plants I could name. Goldenrod has a boldness I've always appreciated, especially because the flowers appear late in the summer, when colder weather is just around the corner and most other wildflowers have faded.

Showy goldenrod is among 125 species of goldenrod, and most of them make excellent meadow plants. (Except for Canada goldenrod, which is aggressive! Avoid it unless it's the only plant you want to see in your meadow.) It has side stalks that are erect or curved upward, while most other goldenrod flower stems curve outward or down.

The species has deep fibrous rhizomatous roots, which makes it helpful in stabilizing hills. It's used to restore prairies, establish pollinator plantings, and shore up riverbanks. *Solidago rugosa,* also called wrinkleleaf or rough-stemmed goldenrod, has many of the same qualities.

Goldenrod species often get blamed for the effects of hay fever, but the real culprit is ragweed, which blooms at the same time. Goldenrod's pollen is relatively heavy and carried away by insects, not the wind.

WHERE IT THRIVES

Resilience. Showy goldenrod can handle long stretches without water, like most goldenrod species. This family of plants is pretty rugged overall—I've even seen another species, seaside goldenrod, growing out of little more than dry cracks in rocks on the Maine coast.

Regional compatibility. Showy goldenrod is native in the United States everywhere from the Rockies east. Every state has a native goldenrod species.

Soil conditions and sun. Most goldenrod thrives in a range of conditions, including dry and clay soils. This plant prefers full sun. It can survive in shade, but won't flower.

PLANTING AND GROWING

Showy goldenrod does well when seeded into a well-cleared site, and it self-seeds easily, but not aggressively. In moist, fertile soils it can spread rapidly and flop over due to weak, rapid growth. You can curb runaway growth in traditional meadows by including plants of similar height in your mix. In meadow gardens, you can limit the spread by digging out the stems and separating them into individual transplants, or just composting them.

Showy goldenrod faces few pest or disease problems, except for the occasional case of powdery mildew.

SMOOTH BLUE ASTER

Symphyotrichum laeve
Height: 2 to 3 feet
Hardiness zones: 3 to 8
Bloom time: August, September, October

Smooth blue aster has a robust root system, which explains its impressive resilience. It grows almost anywhere, including in cultivated gardens, moist woods, dry prairies, and city lots. This exceptional hardiness makes it a good choice for prairie restoration and roadside revegetation projects. But I often recommend it because it's a late bloomer that continues to feed pollinators while other flowers are disappearing. When I use asters in meadow gardens, I favor lower-growing ones, like this one.

WHERE IT THRIVES

Resilience. Smooth blue aster can tap into its well-developed root system to weather drought. It can also resist frost and sometimes blooms into November.

Regional compatibility. Native in all regions of the US, smooth blue aster has many cousins that are also perfect for meadows, which explains why I often use more than one type of aster in a single design.

Soil conditions and shade. This rugged little plant can grow in sandy or rocky soil, in loamy, fertile soil, and in moist, poorly drained soil, as long as it's not wet all the time. Ideally, it needs some moisture and will become somewhat stunted if conditions remain too dry for too long. This type of aster thrives in full or partial sun.

PLANTING AND GROWING

Plugs planted in early spring often mature quickly and flower by late summer. Deer will nibble on this plant, but except in the smallest meadows and meadow gardens, it's vigorous enough to take this in stride.

WILD BERGAMOT

Monarda fistulosa
Height: 2 to 4 feet
Hardiness zones: 3 to 9
Bloom time: July, August, September

I've planted wild bergamot in a lot of gardens over the years with disappointing results. So why am I including it here? Because I eventually realized this short-lived plant's real home is in less-curated spaces—like meadows. In a meadow planted from seed, wild bergamot can come and go as it pleases, shifting locations over the years as it reseeds itself. In a carefully designed garden, this sort of behavior can be a bit unruly.

Wild bergamot's pink and lavender flowers are a magnet for pollinators. The species name, *fistulosa,* means "hollow like a pipe" and refers to the blossom's tube-like shape, which attracts long-tongued bees and butterflies, as well as hummingbirds. The flowers appear in midsummer and bloom for many weeks. After the last flowers die the seed heads remain vertical for months, adding height and texture to a winter field.

Bergamot's close cousin bee balm (*Monarda didyma*) is another plant I use regularly. It has similarly shaped flowers, but with a bright red color. Both plants are members of the mint family, and have fragrant and minty-tasting leaves. This partially explains a common nickname for wild bergamot: horsemint. But horses don't like it, and neither do most animals.

WHERE IT THRIVES

Resilience. The flower can handle heavy rain and drought, but only for short periods.

Regional compatibility. Wild bergamot is best suited to dry climates and air, but is native to almost every state. Even so, it's classified as a weed in Nebraska, which points to the arbitrariness of what gets called a weed.

Soil conditions and sun. Grows best in dry to moist, organically rich, well-drained soil, but it can thrive in less rich soils like clay and sandy loam as well. Although it can handle dappled shade, wild bergamot needs lots of sunlight to produce flowers.

PLANTING AND GROWING

This plant is useful if you've decided to seed into existing grass. Its small seeds easily find their way through grass and leaf litter to make soil contact. Once they do, they typically sprout readily. If you opt for plugs, you can plant them just about any time during the growing season.

In moist conditions wild bergamot is vulnerable to powdery mildew and rust, which can make it look unkempt. In a smaller meadow, you may want to cut it out so your meadow looks better. Thanks to its minty flavor, herbivores want nothing to do with it.

WILD QUININE

Parthenium integrifolium
Height: 2 to 3 feet
Hardiness zones: 4 to 8
Bloom time: July, August, September

From a distance, wild quinine creates beautiful splashes of white. Up close, the flower looks like a miniature cauliflower. It grows on rocky hillsides and in heavy, dry soils, and can adapt easily to richer garden environments. Wild quinine is a good choice if you're working with a hot, sunny site.

Quinine can bloom for up to eight weeks at a time, and bees and other beneficial insects love its pearly white blossoms. Since it's less showy than many other meadow flowers, it mixes in well with eye-popping flowers like butterfly weed, and can be planted in cloud-like drifts among grasses.

The plant contains quinine, which can be used to treat burns. During World War I, the US Army harnessed the plant's medicinal power as a treatment for malaria. More recently, the National Institutes of Health identified *Parthenolide*, a chemical found in wild quinine, as a potential anti-inflammatory active agent.

WHERE IT THRIVES

Resilience. This plant grows a deep taproot, which helps it survive dry prairies and drought-like conditions.

Regional compatibility. It's native to most eastern states, except New Jersey. Across the Hudson River, though, wild quinine is thriving in Manhattan's famous High Line.

Soil conditions and sun. Wild quinine performs to its fullest potential in fertile, loamy soil. It can also survive quite well in sandy, rocky, or clay soils, it just won't fill out as much. This plant prefers full sun but can tolerate dappled or intermittent shade.

PLANTING AND GROWING

Quinine grows easily from seed. Once established, it's able to spread underground, but not aggressively so. Flowers appear in late summer and often last until the first frost.

Too much or too little water will lead to fewer, scraggly blooms. In smaller meadows, if you don't like the look of the dead central stalks in the winter, go ahead and prune them back. The plants are pest-resistant and deer and other herbivores don't care for the bitter, rough leaves.

Meadow plant preferences

Meadow plants are pretty adaptable. Even so, each one has its own particular set of likes and dislikes. Match your plant to your meadow site, and it will flourish for many seasons.

Perennials	Dry	Average	Moist	Sandy	Clay	Loam	Spacing	Zones
Anise hyssop	x	x	x	x	x	x	12–15"	4 – 8
Black-eyed Susan	x	x		x	x	x	12–15"	3 – 9
Blue grama grass	x	x		x	x	x	12–15"	3 – 10
Butterfly weed	x	x		x	x	x	12"	3 – 9
Early sunflower	x	x	x	x	x	x	15–18"	3 – 8
Foxglove beardtongue		x	x	x	x	x	12–15"	3 – 8
Great blue lobelia		x	x		x	x	12"	4 – 9
Hoary verbena	x	x		x	x	x	15–18"	3 – 8
Lanceleaf coreopsis	x	x	x	x	x	x	12–15"	4 – 9
Lupine	x	x	x	x	x	x	15–18"	3 – 8
Meadow blazing star	x	x	x	x	x	x	12–15"	3 – 8
Mountain mint	x	x	x	x	x	x	18"+	3 – 7
Obedient plant		x	x	x	x	x	12–15"	3 – 9
Purple coneflower	x	x	x	x	x	x	12–15"	3 – 9
Purple needle grass	x	x		x	x	x	12–15"	5 – 10
Showy goldenrod	x	x	x	x	x	x	15"	3 – 8
Smooth blue aster	x	x		x	x	x	12–15"	3 – 8
Switchgrass	x	x	x	x	x	x	12–15"	3 – 9
Tufted hair grass		x	x	x	x	x	12–15"	3 – 8
Wild bergamot	x	x			x	x	15–18"	3 – 9
Wild quinine	x	x		x	x	x	15–18"	4 – 8

PREP WORK

A **few years ago Hampshire College decided to turn its** main quad into a meadow. The large, flat lawn gets plenty of full sun and is neither too wet nor too dry. In other words, perfect meadow material. The project was assigned to the college's facility department, which had no experience with meadow design or construction. But they had a job to do and soldiered on, tilling the space once and then planting it with perennial seeds.

Unfortunately, they skipped a few steps. The crew never scheduled a second tilling to destroy the remaining weeds, nor did it find a way to protect the bare ground against new weeds. Soon enough, a swarm of fast-growing, ferocious weeds, reinvigorated by the tilling process, rose up and took over the site.

The college immediately started fielding complaints, which only intensified once the students returned in the fall. While the meadow might have grown in given enough time, the college didn't feel it could take the chance. So a hasty decision was made to scrap the first attempt and replant the space. The college sought guidance from meadow experts, who advised tilling more than once and protecting the freshly cleared site with a nurse crop of fast-growing grass. This time there were no complaints, and the still emerging meadow looks like it will be around for a good, long while.

As anyone who has grown a successful meadow knows, managing weeds is how you prepare a site for planting. Ideally, this means clearing your site of all vegetation, which may seem ruthless, especially if you like some of the plants. But starting with a blank slate gives you the most control over the outcome, and your seeds and plants the best chance of growing up. It also means fewer unwanted weeds down the road.

If you'd rather not scrub your site clean, that's okay. As mentioned, I've had success seeding into existing lawns. Still, if you go that route, be prepared to roll with whatever sprouts, including more of the original plants and possibly plenty of weeds, too.

To clear land, many meadow designers and landscapers recommend using herbicides. They like that they're fast and allow them to start seeding almost immediately. But herbicides are not always as effective as their devotees claim. While they can kill most plant life, they won't get rid of the seeds underground, which means weeds can easily resprout and require a second round of spraying.

Over the many years I've spent prepping sites, I've never once used a weed killer. I learned early on that the effects can be far-reaching, unpredictable, and deeply harmful to the environment. Simply stated, using herbicides is incompatible with the goal of supporting pollinators and creating regenerative habitats. Fortunately, there are safe and effective alternatives.

TURN OVER YOUR LAWN

Tilling the earth by turning over weeds and grass so they die is how I clear land most of the time. The method works even when digging up all but the most chemically contaminated lawns because it mixes organic matter back into the soil, and jump-starts the soil's recovery process. Once cleared, tilling also prevents weeds from coming back, although not all at once.

Seeds often won't sprout unless they've been jostled, exposed to cold and then heat, watered, or otherwise disturbed. Tilling flushes out unwanted weeds by disturbing seeds lying dormant in the soil. Roots often start growing back, too. Depending on the equipment and the tenacity of your existing plants, you'll likely need two or three cycles of tilling to wipe out weeds and weed seeds, with ten to twenty days in between cycles. During those between times you can see what, if anything, is still growing. Once about 95 percent of your site is sprout-free, it's ready for planting.

Tilling also works if your soil is too hard-packed to grow in. If you can't sink a shovel without using force, you have heavily compacted soil. Tilling the soil breaks it up and allows in the water and air that supports microbial health. Loose soil also makes it easier for seeds to settle in and take root.

Tilling does burn gas, and disturbing the soil releases small amounts of carbon into the atmosphere. When used regularly over the course of years and decades, as in conventional farming, it can significantly deplete the soil of carbon. But when you're turning over your soil only a few times to create a meadow, the impact will be more than offset by all the carbon your meadow will capture over its lifetime.

How to: Tillers are machines that use spinning tines to chew up grass and weeds. For small to medium-sized jobs, my go-to machine is a walk-behind hydraulic rototiller. If you're a DIY type, walk-behind hydraulic tillers are fairly easy to rent, but they can be gnarly to operate. It might just be wiser to sub out this particular job to a landscape company, especially if your site is large, which can call for a plow and harrow.

On very small sites you can use a garden tiller, basically a small rototiller, though it doesn't do well breaking up dense turf. I tend to use hand tools—like a shovel, a mattock, which is similar to a pickaxe, and a rake. I'm not going to lie, this is a lot of work. With lawn grass, I use a flat shovel and scalp it. But the real chore is digging out weeds by their roots. Still, while demanding, this approach has the least impact on the environment, and you only have to do it once.

SCORCH YOUR WEEDS

To clear small meadow sites I sometimes recommend solarization, which smothers and scorches existing plants by covering them with sheets of heavy black plastic. The plastic deprives grass and weeds of life-sustaining sunlight, but it's the intensity of the heat that does the real work. It scorches the plants, cooking the uppermost part of the root system, and kills seeds close to the soil's surface. The heat also knocks out healthy soil organisms in the top half inch of soil, but since the beneficial bacteria and microbes a few inches down aren't affected they're able to quickly make a comeback.

The solarization process usually takes about four to six weeks. While I'm not crazy about using plastic, you can reduce your impact by borrowing a sheet from someone you know, or getting it on CraigsList, and then passing it along once you're done with it.

How to: Solarization only works when the plastic is somewhat opaque, black, and flush with the ground. Before starting, mow your grass as short as possible. Since heat is key to making this process work, cover your site just before the hottest part of the summer. Weigh it down using stones, soil, or anything else that won't blow or wash away, and keep it in place for a minimum of four weeks.

When the soil underneath the plastic is bare, test it to see if it's ready for planting. Dig a few holes, at least four to eight inches deep, and if you can't find any living roots that are sprouting shoots the sun has done its job. Otherwise, cover up your site a little longer. Solarization is an effective way to thoroughly rid yards and small fields of unwanted vegetation without using chemicals or machinery, as long as you don't mind looking at a vast sheet of heavy black plastic for awhile.

PLANTING GUIDE

As a young landscape designer just starting out, one of my first projects was redesigning and renovating the grounds at Robert Frost's former home in Franconia, New Hampshire. Frost was a deeply devoted amateur botanist and orchardist. In fact, he moved to this property with a grand view overlooking the White Mountains to grow apples. In homage to his abiding passion for pomology and botany, one of the design goals for the seven-acre property was to plant an apple orchard, as well as a meadow.

To prepare for the meadow plants and apple trees, we needed to restore a one-acre area that had been meadow going back to the early twentieth century, when Frost lived there. We started by cutting back and clearing out all the woody brush that had recently begun to proliferate. Then we planted forty heirloom apple trees. We didn't entirely clear the site of weeds and grass, mostly because I was still getting the hang of meadow building.

To make the space more visually attractive and biodiverse, I chose natives that would do well in the hillside's moist conditions. Then I organized a planting day for volunteers. Using trowels and digging knives, we carefully dug holes for each of the perennial grass and flower plugs, and planted them into the reclaimed meadow.

This was my first time attempting such a feat. In hindsight, I did almost everything right—except for one thing. I failed to mark the exact location of each new plant. In almost no time, the existing weeds and grass grew up around those little plugs, which hid them from view and made it hard to keep them weeded. By the end of the summer, they had all but disappeared.

But not all was lost with the Frost meadow. Along with the plugs, I'd sprinkled in some native perennial seeds and over the next few years those seeds germinated and grew. The lupine was especially happy there, and on my occasional return trips I've watched the purple spires spread more every year.

Since then, I've learned that the practices for building a meadow are fairly universal, no matter where you live. Still, there's no single right way to plant one. Even meadow experts can disagree on tactics. I used to puzzle over this but now I compare it to using different recipes to make the same dish; any discrepancies in the amount of olive oil to use, or whether to go with thyme or tarragon, are just variations on the same theme. When growing meadows, whatever gets results is what works.

PLANTING SEEDS

Starting from seed is the ideal way to plant a large meadow because you can cover a lot of ground at relatively low cost. Meadows grown from seed typically take two to four years to mature, although you may see results sooner. When planting from seed, you won't need to create a planting plan or an on-the-ground design; all you have to do is distribute a mix of seeds over your site, preferably a cleared one, and nature will sort out what thrives where. As long as the species in the mix match the site conditions, this approach typically gets excellent results.

Decide what time of year to plant. You can plant seeds almost any time of year, including winter, because they won't sprout unless they find the conditions suitable. If the temperature or rainfall levels aren't to their liking, they'll just lie dormant, waiting for their moment. That said, you probably want them to sprout as quickly as possible, not only because it looks better but to prevent your site from becoming overrun by weeds.

With this in mind, aim to plant your meadow in the late fall or early spring—either time is fine. In colder climates, the second half of the fall is preferable because cold-tolerant plants often need to go through a winter before sprouting. If you don't hit these targets precisely, no worries. Just about any time in the spring or fall can work.

Use a nurse crop. Nurse crops, so-called because they protect new seeds and seedlings, typically consist of one annual species, like ryegrass, oats, or barnyard grass, spread evenly throughout a meadow. Because they sprout quickly, nurse crops protect the soil against erosion and sun. They also dress up your plot with fresh green growth until your perennials can make an appearance. Without an annual nurse crop, your new meadow would look an awful lot like abandoned dirt.

But the best reason to plant a nurse crop is that even the most energetic site-clearing effort won't prevent weeds from making a comeback, thanks to birds making flyovers, gusts of wind bearing seeds, and weed seeds you didn't catch the first time. A nurse crop blocks out most weeds by filling out the open space and taking up the available sunlight, which gives meadow plants time to establish.

Something else to appreciate about these plants is that they're temporary. By the second season, around the time your perennial grasses and faster-growing meadow flowers finally show up in force, they'll be gone.

Find a good seed supplier. Reputable sellers don't sell species that can be invasive, like purple loosestrife or knapweed. Instead, they focus mostly, or exclusively, on selling native plants. Good seed suppliers also keep track of viability rates, that is, the likelihood seeds will germinate. Each batch varies to some degree since they are all harvested, processed, and stored under different circumstances. I also seek out suppliers who avoid using synthetic fertilizers or pesticides, and make their environmental practices clear.

If you can't find a good local seed source, consult with a national provider that can help you choose a list of the native species most likely to thrive in your locale. Since costs and availability can vary quite a bit, I tend to ask for estimates from more than one supplier.

Order your seeds. Seeds are sold by weight. Each species has different-size seeds and germination rates can vary enormously, so figuring out the right blend of plants can involve professional-level botany skills. Because these factors are way too complicated to track, I rely on my seed suppliers for help. I share with them the total area of the site, the list of plants I've chosen, the percentage of grasses I'd like to see, and the species I want to highlight, or play down. I always order an extra 10 or 15 percent in case some seeds don't pull through.

Sow your seeds. You can sow your seeds one species at a time, though this is pretty tedious. I either mix them all together, or ask my suppliers to sort seeds by size into three mixes—small, medium, and large. What's most important is that you spread your seeds evenly. To keep them from clumping together, I mix my seeds with a medium like sawdust, coco coir, or sand, moistened lightly so the seeds stick to it.

I tend to sow the nurse crop seeds just before the perennials so I can see where I've spread them, and make sure the ground is well covered. This is obviously easier with larger seeds like annual ryegrass. If you want to just mix your nurse crop in with the other meadow seeds, or spread them afterward, that's fine, too.

With almost any size field, I divide the area into sections. Working in smaller zones makes it easier to cover the whole field more or less evenly, without running out of seed. For larger spaces, I often use a hand-pushed broadcast spreader. For small meadows, I just spread seeds from a bucket. If you live in a windy area or worry about the birds getting them, lightly rake seeds into the soil no deeper than an eighth of an inch.

Once you've spread your seeds, you're done. There's no need to water the area unless you live in an arid zone, like parts of California.

In that case, an occasional light sprinkling for the first two or three weeks can help trigger germination.

PLANTING PLUGS

Plugs take a little more effort because you have to dig holes and add a layer of mulch, but it's still light work and the fastest way to grow a meadow. The plants can become established in the course of a single growing season if they get enough sun and water. If you're accustomed to growing annuals, this may still require an adjustment in expectations—something I learned years ago when I was hired to design a meadow garden at the entrance of a very large Air Force base.

The grounds crew had always planted annuals, which needed lots of care, and a forward-thinking natural resources manager wanted to plant something more sustainable. We worked together to come up with a new design and, since the site was small, decided to use plugs. We planted them on the day of a large airshow, working mostly without talking because the jets buzzed so loudly. I remember hoping this very public garden might inspire others to plant their own meadows.

But it never had the chance. The following May, a high-ranking officer, unhappy that the meadow had yet to produce any color, ordered that all the plugs be ripped out and replaced with the same old annuals. Had he waited only another two or three months, he'd have seen the gorgeous meadow we planned.

Decide what time of year to plant. Spring generally works best. The ground tends to be moist and plugs start growing immediately without much, if any, watering. By the time the weather turns hot, they're usually strong enough to keep growing with little or no help from you. But it's also fine to plant plugs in the fall, unless you're working

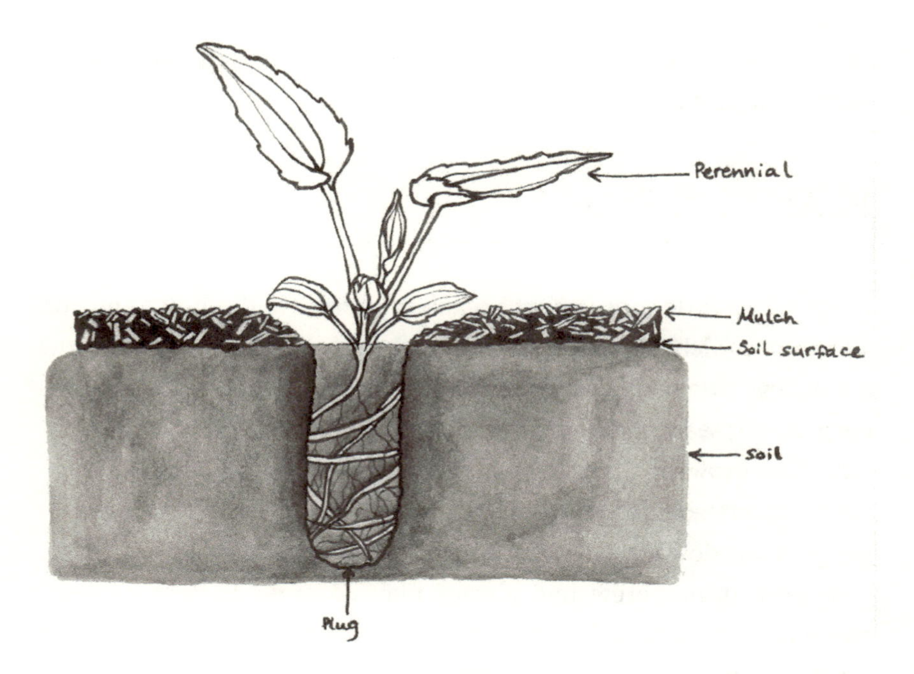

Dig a hole the size of your plug's rootball, and lower in the plug. Fill in the hole, tamping with your fingers to eliminate air pockets. Make sure the bottom of the plant, where it meets the top of the plug rootball, is flush with the ground. Spread one and a half inches of mulch around each plant, but taper it off so it doesn't touch the plug base, and promote rot.

with plants that prefer dry conditions, which tend not to appreciate long, wet winters. Young, dry-loving plugs do better in the spring.

You can plant plugs in the summer but they'll need more watering; in direct sun, baby plants can dry out in a matter of hours, or even faster. Since regular watering can stimulate weed growth, a summer planting means more work for you.

Figure out how many plugs you need. Many nurseries don't sell plugs, so I usually go online to find them. They're typically packaged in trays holding thirty-eight to fifty plants and some suppliers allow for mixing and matching, which is helpful if you're working with a very small space or have a tight budget. Unlike planting from seeds, you don't need to buy extra plugs.

To figure out how many plugs to buy, research how much space each species needs when full grown. Nurseries almost always supply this information. Then, do the math to see how many you'll need to fill your site. Order your plugs for delivery only when you're ready to plant. Plugs don't do well sitting in trays for very long.

Lay them out before planting. If you mapped a design on paper, now's the time to use it (see page 43). Make sure to lay out your grasses first, and then blend in the flowers. Once you've organized your plugs, adjust the spacing as needed, about fifteen to eighteen inches apart will usually do. But plants with a larger footprint can be set as far apart as twenty-four inches while more diminutive plants can be closer together. Ideally, the perimeter of each mature plant should touch the shoulders of its neighboring plants so it's harder for weeds to move in once your meadow garden fills out.

To plant them, use a trowel or digging knife to create a hole the same depth and width as the root ball, and place the plug so the base of the plant, where the stem meets the soil, sits flush with the ground.

Water as needed. Baby plants have very small root systems, which means they're vulnerable to drying out until the roots have grown long enough to find moisture on their own. In the absence of rain, water your new transplants by hand or with sprinklers for the first couple of weeks—longer if planting during the heat of the summer.

Keep an eye on your new plugs the way you would new chicks, and water only as needed. You don't want your plugs to become dependent on watering; you want them to toughen up and adapt to site conditions. At the same time, you don't want them to wilt or completely dry out. You'll know it's time to water your plants if you stick your finger in the soil and it's dry up to your first knuckle. Once your plants start growing and they've taken root, you can taper off. Plugs can often be weaned off watering in a matter of weeks.

Apply mulch. Since there is no nurse crop for plugs, use mulch to keep away the weeds until they can grow in. I use a natural bark mulch (preferably partially decomposed), but you can use almost any type of organic mulch, including dry shredded leaves, untreated grass clippings, and straw. (Don't use hay, which has seeds. Or wood chips! They can temporarily leach nutrients out of the soil and stunt your plants.) Because it holds water easily, don't mulch in arid or dry meadow gardens.

To spread mulch, I shovel it into piles between the plugs and spread it out by hand, about an inch and a half deep. Be careful not to smother the base of the plants where they meet the ground or they may develop fungal growth and rot.

PLANTING IN UNCLEARED SOIL

The most straightforward approach for turning lawn into meadow is to simply let your lawn grow out, and add perennials. This technique is appealing to people who'd rather not create a temporary bare patch. It lets you skip out of using machinery, plastic, or herbicides. It also happens to be the fastest and least expensive route to turning a lawn into a meadow.

Though this approach can work surprisingly well, the outcome is more of a gamble. The grass species used to grow turf are not typically optimal for meadows. When fully grown, they can fall over and end up a mishmash of heights. Lawn grasses also tend to be cool-season grasses, which means they spread thickly, forming a dense mat that can inhibit other meadow species.

All that said, when we had no choice but to seed into the old lawn at The Carle, it worked out pretty well. I recommend choosing plants that are particularly robust so they can power through the neighboring weeds and grass.

Pick a time to plant. You can plant seeds into a lawn at any time, whether or not your site is cleared, and almost all the same factors come into play. But if you're working with plugs, the optimum time is early fall when weed growth has slowed, with spring a close second. In summer, existing grass and weeds are more likely to outgrow and crowd out your new plants.

Mow one last time. This will be the last time you mow your yard for anything other than annual maintenance. To give your seeds and plugs a fighting chance, mow as close to the ground as the mower settings allow. If you're left with extra thick grass clippings that might block your seeds, rake them up and compost them. With plugs, go ahead and leave the clippings as mulch.

Plant your plugs. If you're using plugs and seeds, add the plugs first. Planting plugs disturbs the soil, which can end up burying seeds already on the ground. I space plugs further apart in uncleared sites since the ground is already covered, and organize them into clusters for a natural look. After planting, clear a few inches of space around each plug to give it room to grow, and mark it with a stick or little flag, so it's easy to find when you're weeding and doing basic upkeep. If watered, as needed, and weeded, your plugs will fairly quickly become part of your meadow matrix.

Spread your seeds. Scattering seeds on top of a lawn can seem like a lost cause, but there's a good chance many will germinate. You can increase your odds by sowing a greater variety of species and, of course, using more than the recommended amount. It can help to use a drill seeder, though it's not a requirement.

Drill seeders slice into turf to bury and then cover each seed. Most drill seeders are pulled by tractors but there are smaller, walk-behind versions, too. The best drill seeders have different compartments for seeds of varying sizes, so if you use one ask your seed supplier to sort your seeds by size (small, medium, large). Plant seeds very close to the surface of the soil—no more than one-quarter inch deep—so they get light immediately after sprouting. Note: If you use a drill seeder, plant the plugs after you seed so they're not chewed up by the machine.

Meadow Tools

Broadcast seeder
Distributes seeds evenly.

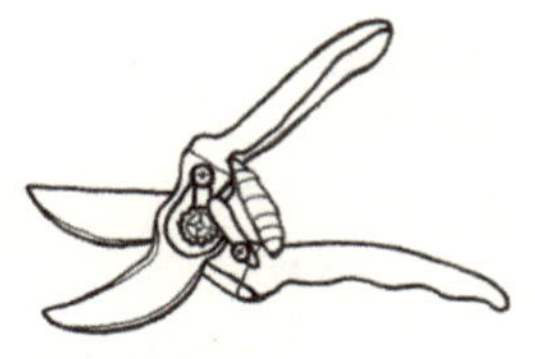

Clippers
Strong enough to cut back woody plants.

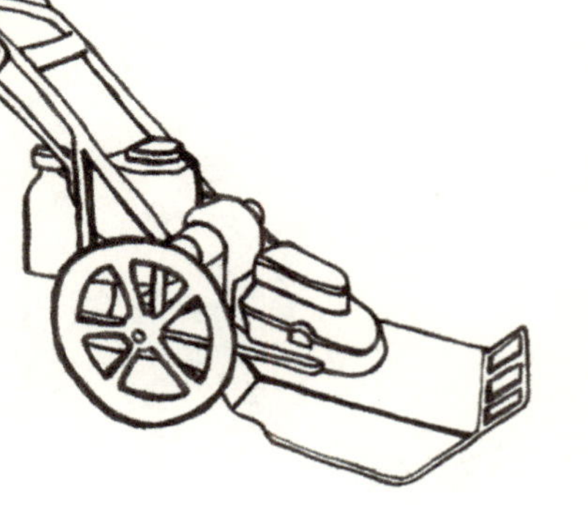

Brush hog
Mows lawn and is rugged enough to chew up woody plants, making it especially useful in rough terrain.

Digging knife
Digs holes for planting plugs. Helps extract weeds from your meadow garden.

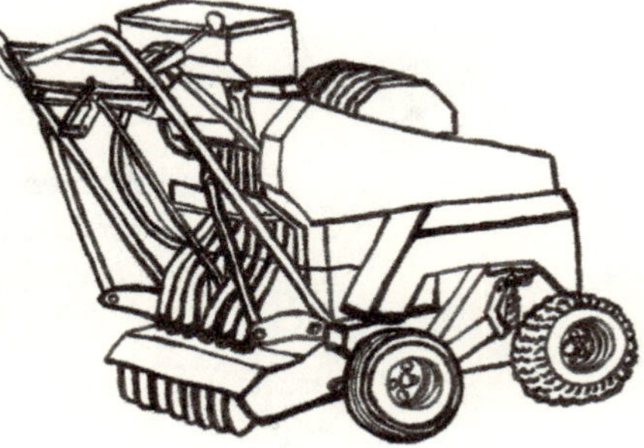

Drill seeder
Scores seeds into the ground, which is useful when seeding into existing turf. This walk-behind version is suitable for small yards.

Hoe
Helps dig out small patches of weeds after tilling, without straining your back.

Scythe
Selectively cuts weeds or mows larger swaths of meadow

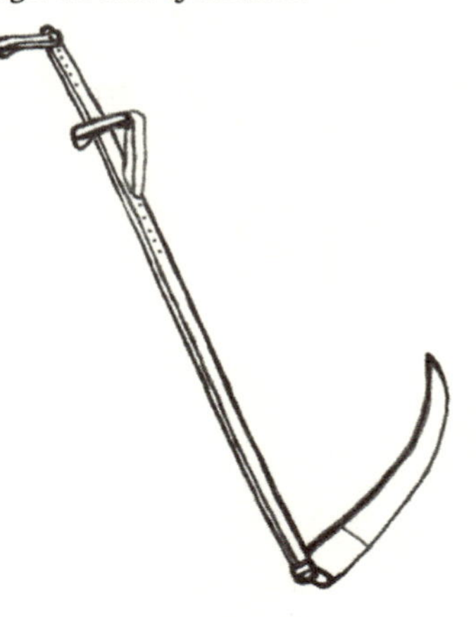

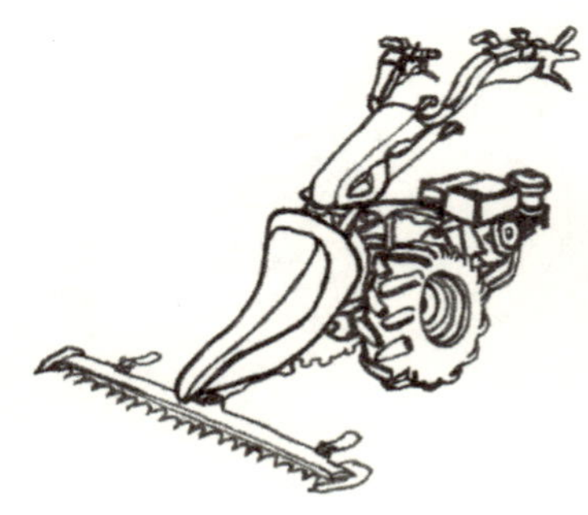

Sickle Bar Mower
Cleanly scissors vegetation without chewing it up, which protects the insects and their eggs sheltering in grasses.

Trowel
Useful for planting plugs and deep-rooted plants.

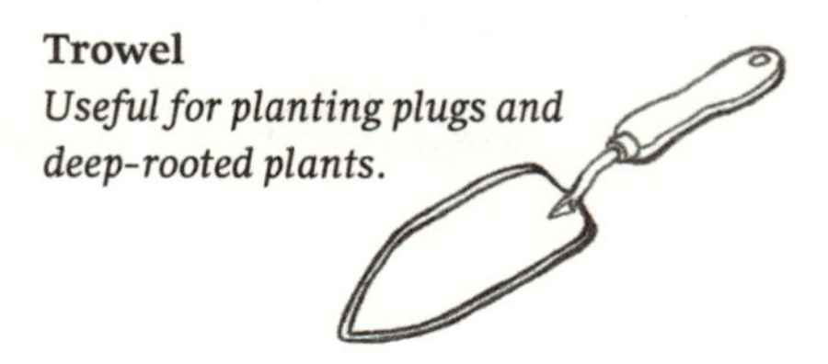

Weed Wacker
Spot-clears weeds. Especially useful for sloped areas.

UPKEEP

A meadow garden I worked on many years ago was on a steep slope facing a well-travelled road in Holyoke, Massachusetts. When I arrived in the spring to evaluate the site, I found a drab, overgrown embankment filled with non-native, woody plants like honeysuckle, barberry, bittersweet, and multiflora rose. The presence of all those unruly plants gave the house behind the hill a derelict feel.

The homeowners asked me to clear out the mess and fill the space with as much color and as many flowering species as possible. So I designed a meadow garden that would flower profusely and thrive with little to no attention—purple coneflower, wild bergamot, butterfly weed, anise hyssop, asters, bee balm, switchgrass, and more. Once they'd greenlighted the plan, my team and I began the process of removing the existing growth.

Using clippers, loppers, and shovels over the course of a full day, we painstakingly cut back the woody plants and dug out every last root. The next day, we returned to plant hundreds of plugs, taking care to space them at just the right distance so they'd fill in properly and leave little room for weeds. Once the plants were in the ground we mulched the slope, gently watered the new plugs, and the job was done.

Before leaving I told my client to keep the plugs from drying out for the next couple of weeks or so, and pull any weeds. From there on out, all he'd need to do to keep his new meadow going was to mow it each spring. He assured me he was on it, and in those first few years he was good on his word.

I enjoyed watching the meadow's slow-motion explosion of color on my occasional drives by the property. The bee balm was the first to take off, and by midsummer its bright red flowers had sprung up all across the slope. Clusters of bright black-eyed Susans soon joined in, and eventually asters, which created a patch of striking purple that lasted until well after the first frost.

The whole show repeated itself the following season, and the next. But when I drove by the following spring, I noticed a few spiny multiflora rose and aggressively vining bittersweet, a sign the embankment had not been mowed. Even healthy native flowers would be no match for these aggressive intruders.

To the untrained eye, it looked like nothing was wrong. The species we'd planted still flowered that summer, but their days were numbered. If the encroaching woody plants weren't cut back they'd completely overwhelm the meadow, and that is exactly what happened. The following year I watched as the brush took over the meadow, growing taller than the meadow plants and starving them of sunshine. The ones that remained suffered, and produced far fewer leaves and flowers.

My relationship with clients typically lasts only until I know the landscape is well established, so I don't know why the meadow owner suddenly stopped tending it. Perhaps he moved, or grew ill. Maybe he just lost interest. Either way, it took only one more growing season for the meadow to be completely swallowed up by shrubs and taller plants.

Even then, the meadow could have made a comeback had the brush been cleared because there were probably a few little meadow plants still hanging around. But it was not meant to be. Today, there's no sign that a beautiful meadow garden once flourished there.

A meadow is defined as an open space populated with grasses and herbaceous plants. Trees and shrubs are not part of the equation, which means that for a meadow to last there must be some intervention—like fire, mowing, or grazing animals—that keeps woody plants from taking over. A lack of water can also keep woody plants in check, which is why wild meadows tend to last longer in the Great Plains, parts of California, and places in between. But meadows do need a minimum level of moisture to stay alive; below that threshold, meadow ecosystems shift toward desert.

With enough water and nothing to get in their way, it's only a matter of a few years before trees and bushes encroach on and subsume a meadow. Fortunately, this is easy to avoid since meadows need very little help from you to endure.

CARING FOR A YOUNG MEADOW

It's rare that I forget a landscape I've designed. Like friends who are out of sight but never far from mind, each one stays with me. I feel a deep connection to my sites, and work hard to come up with a creative design for each one. But the primary reason I remain close to my meadows is because the true gauge of their success is how well they develop over time. It can take a low-maintenance meadow two to four years to become well established and self-sustaining. Until then, it sometimes needs a little help.

Still, don't fuss over it. Every meadow, no matter how it was planted, grows better if it isn't coddled. Don't water a meadow unless you're using plugs, and then only until they root. And don't ever add fertilizers. Forcing your plants to dig deep for water and nutrients will make them much stronger and better able to handle extreme storms, heat waves, and drought. Not surprisingly, this hands-off approach makes maintaining a meadow, even a developing one, pretty straightforward.

Weed your new plants. No matter how well you've cleared your site, you will get some weeds. But meadows planted from seed sprout so many seedlings they'll end up crowding out most weeds. Since seeded meadows tend to be larger, they can also visually absorb the presence of weeds. Besides, a few dandelions, plantains, or crabgrass plants are no big deal.

When seeding into a lawn, you'll need to do more weeding. In the first year after planting The Carle meadow, we went in with a scythe every ten to fourteen days to cut weeds before they had a chance to blossom and seed. (By the way, *it's really important to prevent weeds from going to seed!*) We cut the weeds to a height of about six inches to protect the short, young meadow plants.

For a meadow garden just getting started with plugs, pull weeds when you see them—unless you're getting crushed by them. If that's the case, cut them back. You can do this using hand tools like clippers, or grass shears. In a larger space you'll want to use a weed wacker or scythe.

Mow as needed. A young meadow can be mowed as many as three or four times, especially in its first year, if it means keeping perennial weeds from going to seed. But reserve it as a last resort. Use a high setting—six inches is good—so your perennials continue to grow. Leave the cut stalks and leaves right where they fall, to act as a mulch and add organic matter to the soil.

Practice pest prevention. A mature meadow can eventually support all kinds of life. But when it's just getting started you'll want to protect it from herbivores who treat it like a buffet. Meadows planted from seed can handle some grazing right from the start, unless it's really heavy. Small meadows planted from plugs are more likely to need protection. Deer, rabbits, and other animals typically leave clues, like tracks, scat, or chewed plants, before doing any real damage, which gives you time to use an organic pest deterrent. (See "Organic pest controls," page 125.)

CARING FOR A MATURE MEADOW

Once a meadow's grasses and flowering species have reached their full height, it needs very little care. This usually happens by the second year for meadow gardens and can take two, three, or even four years for meadows grown from seed.

Mow your meadow every year. The one minimum maintenance requirement for every meadow, no matter how old, is a yearly mowing or, if it makes sense, a controlled burn. Woody invaders are slower growing than meadow plants, which means that as long as you cut them back annually, your meadow plants should be able to shade them out while getting the sun they need.

You can mow in late fall, but I prefer early spring. Leaving meadow plants standing over the winter supplies seeds for animals and birds when food is sparse, as well as shelter. Many pollinators, like skipper butterflies, overwinter as caterpillars in clumps of dried grass blades, while Eastern-tailed blue caterpillars take refuge in seed pods. Waiting to mow until spring allows these insects and their eggs to survive the winter, protecting them and the other life they help support. It's also the reason a meadow can look lovely in February.

Aim for a height between four inches and eight inches, and don't mow during the growing season, which would demolish your perennial grasses and flowers.

Consider doing a controlled burn. Another way to manage woody plants and brush is through controlled burning. Fire is the single most effective way to naturally maintain a meadow over time. When it burns through a meadow, it usually advances as a line that lays waste to dry vegetation, including dead and matted plants. Since grasses and meadow plants don't create a lot of heat when they burn, the plants and seeds just below the surface are able to regenerate with renewed vigor.

For obvious reasons, an annual burn is not a practice that works well in densely built neighborhoods, and certainly not without the oversight of trained professionals. But in locations where there's little risk of damaging property and burns can be easily controlled, an annual burning can be a nice alternative to mowing.

I've never had the chance to manage a burn myself because the meadows I create are usually too close to structures and woods, and I haven't received the necessary training. If you're considering this option, be sure to consult with an expert.

Manage pests naturally. A healthy, mature meadow is home to many creatures—and you may not be crazy about having all of them in your yard. Since your goal is to create a regenerative landscape, team up with nature to manage the pests. If you find more mosquitoes in your yard after planting a meadow, for instance, install a nesting box for tree swallows or martins, and a nice, big bat house. Swallows and martins are insect-catching birds that spend their days eating mosquitos. A single bat typically eats thousands of the insects before roosting. When a mid-winter snow melt revealed that The Carle meadow had become a haven for many hundreds of voles who were gnawing on the mature apple trees, I recommended installing an owl nesting box. Owls are a native natural predator of voles, and a family of these nocturnal hunters can eat a thousand small rodents in a single season.

Organic pest controls

A meadow attracts a lot of life and can usually handle it just fine. But in certain situations, especially in small meadows and meadow gardens, it can be useful to have a few organic pest controls on hand.

Pest	Problem	Solution
Deer	Deer tend not to be a problem in meadows planted from seed unless their grazing is very heavy just as a meadow is getting established. Meadow gardens are more vulnerable because of the finite number of plants.	Use a plant mister or a pump sprayer to apply a premade solution like Deer Stopper that tastes bad to deer. Some of these can last for weeks and they tend to be very effective. If grazing pressure is strong, you will need to apply the solution more than once.
Woodchucks, marmots, and rabbits	These large rodents can wreak havoc on a very small meadow or a meadow garden. It doesn't usually happen all at once, but once they decide they like some of your plants they will eat them until they are gone.	Castor oil powder is a safe way to discourage them. Be sure to follow the directions of the product you buy, like MoleMax. Repeated applications may be necessary.
Gophers, voles, mice, and small rodents	Small rodents usually aren't a problem, especially in larger meadows. But meadow gardens can suffer if you have too many because they like to eat roots. If your plants appear to be unhappy or are suddenly prone to wilting, look for burrows just under the surface of the soil right at the base of the plant.	Castor oil powder works well as a repellent. These critters just don't like it and will avoid it and anywhere it is spread.
Birds	Birds are not usually a problem. But if you have flocks of them trying to eat your newly seeded meadow it's best to discourage them.	String up bird tape, a reflective ribbon that scares them away. Hawk or owl decoys and other bird-scaring devices can work well, too.

(continues on next page)

Organic pest controls *(continued)*

Pest	Problem	Solution
Slugs and snails	Some snails and slugs consider meadow plants a food source. This is generally only a problem if the weather stays wet so long that they keep on feeding.	There are lots of home solutions people swear by, like garlic oil and spreading coffee grounds around your plants. (Careful, coffee grounds are very acidic!) I recommend stopping these critters with an application of iron phosphate, a naturally occurring mineral found in some slug products.
Ants	Ants are as much a part of meadow ecosystems as the grasses that grow there. They're only worth blocking if they get inside your house.	Diatomaceous earth (DE) is a naturally occurring rock that can be crumbled into a fine powder. When insects come in contact with it, they become dehydrated and die. Scatter it on the ground around the perimeter of your house and ants won't make it inside. Reapply after rain.
Aphids	Aphids form colonies that pierce the cell walls of plants and drink the sap. This weakens the plants and makes them vulnerable to disease and blight. It also makes otherwise beautiful plants look unattractive.	Release native ladybugs into your meadow garden. Ladybugs are a voracious predator of aphids. Or blast the aphids with a hose a few times, an approach that works surprisingly well. Misting them with insecticidal soap also works.
Powdery mildew	This fungus creates a white powdery coating on leaves. It doesn't usually cause major health problems for plants, but it can lead to rot.	Powdered sulfur keeps the problem from spreading further. You can find it in a number of products. Be sure to follow the directions. If it really bothers you, go ahead and remove plants with particularly ugly cases.

COMMUNITY BUILDING

A few years ago, a friend bought a summer cottage on a small island off Portland, Maine in Casco Bay. The house is close to the road and a little too tall for its lot, the former playground of a one-room schoolhouse. It had a lawn and she added shrubs and clusters of flowers to help soften and ground the house, but it never really looked right. She also hated the fumes, noise, and habitat destruction that came with regular mowing.

So a couple of years ago, about halfway through the summer, she mowed the lawn one last time, planted a few small meadow plants, and threw down a perennial meadow seed blend. Soon after, she left town for the winter.

When she returned late the following June the former lawn had grown into a tangle of overgrown grass and weeds. Worse, every time it rained, the water flattened the grasses, which made the house look forgotten. But she also noticed that soft reddish grasses had sprouted while she was away, along with Queen Anne's lace, purple asters, and clusters of vibrant black-eyed Susans. Despite the unruliness of the emerging meadow, the house already seemed less suburban-looking and more suited to its casual island setting.

As she aired out the house and settled in for the summer, she started hearing from local friends. They asked, cautiously, how things were going. One ran up to her husband, threw her arms around him, and asked if he was okay. Where my friend saw the beginning stages of a beautiful, wild meadow, her neighbors saw a lapsed lawn and assumed the worst: illness, divorce, financial problems.

My friend sought to reassure her neighbors, explaining that she was allowing the lawn to become a meadow. Almost every one of them told her they *didn't realize it was intentional!* And with that, people slowly came around to seeing the meadow differently. A few friends told her they liked the change, and as the summer wore on strangers stopped by to introduce themselves and let her know they supported what she and her husband were doing.

Still-skeptical friends were invited to come over, stand on the grass path she kept mowed next to the meadow, and just listen—to the singing of the crickets, the buzzing of different native bees, and the birds calling from neighboring trees. It was a chance to study monarch butterflies and hummingbirds up close, and discover that many tiny ladybugs live on the island.

Simply listening and watching was a practice my friend returned to whenever she had her own doubts. It would take at least another year for the meadow to mature and really look good. But by the end of that first summer, a few of her friends told her they were considering planting meadows of their own.

The level of open-minded acceptance and community warmth my friend enjoyed is unfortunately not everyone's experience. Having worked closely with hundreds of landscape design clients over the years, I've seen firsthand how subjective aesthetic preferences can be, and how widely they can vary. One person's Garden of Eden can be another's out of control, tick-infested nuisance.

For many homeowners, keeping a pristine lawn is a point of pride. This long-held belief that smoothly cut grass should be our default landscape choice has been reinforced by towns, cities, and homeowners' associations across the country that set rigid guidelines for lawn care. Even in unrestricted areas, it's not unusual for neighbors to be concerned if someone lets their lawn go unmowed or get messy.

Sometimes that concern can turn hostile. Every year I hear reports of altercations over lawns and their upkeep. In one of the more high-profile incidents, Senator Rand Paul had six ribs broken when he was attacked by a neighbor angry about Paul piling grass clippings near a shared border. Like I said, some people take their lawns very seriously.

If your neighbors aren't keen on your meadow plans, one strategy is to do as my friend did—just plant your meadow and give them time. The beauty of a healthy, thriving meadow can be hard for even the most obdurate turf-grass aficionados to deny. Nothing deflates the argument that meadows are unattractive better than a profusion of appealing textures and colors that last the entire year. A well-built meadow is its own best selling point.

But I'd be lying if I told you this strategy works for everyone and sometimes, no matter how lovely the meadow, neighbors just don't approve. A few people I've known have forged ahead anyway, meeting the hostility head-on. They've reasoned that sometimes you just have to do what's right for you, regardless of what others think. Perhaps annoying a few neighbors is a small price to pay for helping to reduce our massively outsized carbon footprint on this planet. There's certainly no good reason this country needs an area of mowed grass the size of Washington State.

Still, I wouldn't recommend this approach. You're the one who has to live in your community, and anger isn't fun to come home to. Plus, if the value of your meadow gets lost in local drama, it becomes much more difficult to inspire other people to transform their rarely used or overtreated lawns into meadows of their own.

INTRODUCE YOUR MEADOW TO YOUR NEIGHBORS

Accepting that not everyone will appreciate your meadow can save you a lot of headaches, because it'll force you to get out in front of any problems. Before you begin the meadow-building process, let your friends and neighbors in on why you're taking this step.

Explaining your reasons will go a long way toward helping people understand, and maybe even support, what you're doing. In my experience, most people don't know much about the ecological impact of meadows versus lawns. And while there are those who really like them, most of us are not all that attached to lawns. We're just used to them. To ease your meadow into place, you might consider one of the following approaches.

Start small. If you're anticipating resistance, consider planting a smallish meadow. If all goes well, you can expand it—gradually. When we built the meadow garden next to Northampton City Hall, we started by planting one-eighth of the targeted space to gently introduce the public to the idea. In its first year, local residents saw for themselves that the meadow wouldn't grow too tall, and the city learned how little maintenance it required—a big selling point when we pitched the idea. It didn't take long for the director of city properties to reach out and ask if I'd organize a planting for the rest of the slope. He told me the beauty of this rapidly establishing meadow garden had convinced and inspired the city to continue its support.

Keep it neat. In addition to cutting back weeds occasionally to keep your meadow looking well-groomed, you might also mow a border around it. Neatly framing a meadow signals the new growth is being managed and not a sign of a lawn run amok. Mowing the edges also helps alleviate concerns about ticks by keeping taller grasses and meadow plants away from streets or sidewalks where people walk. Mowing a strip four to eight feet wide around the perimeter is usually enough to create this park-like effect. If you've mowed pathways through your meadow, this outer perimeter can be part of your little trail system.

Hide it. If local regulations or public opinion prevent you from turning your front lawn into a meadow, consider building one where it's less likely to be seen, like your backyard. Even in towns without any prohibitions, it can be useful to practice and hone your meadow-making skills in private before sharing them more publicly.

Another way to hide your meadow from street view is to create a screen of ornamental shrubs. This is something I did recently for a client who lives in a conventional suburban neighborhood. A hedgerow would have worked just fine, but our goal was to create a visual screen that wouldn't grow into a solid presence. So we planted shrubs in small groups, staggering their placement so they'd break up sight lines from the street while leaving space between each island of shrubs. In this way, we were able to create a natural effect in keeping with the look of the meadow.

Raise awareness. People like knowing what's going on next door, so tell them! Honoring their curiosity and concern can go a long way toward opening up people's minds. Whenever I spearhead gardens in public settings, I reach out to everyone who might be affected to explain what we're doing and why. We have presented the plans at town selectboard meetings, posted signs in public explaining the project, emailed neighbors to let them know what's coming, and made ourselves available to answer questions. Even people who are initially suspicious or resistant tend to soften after hearing what you have to say, especially if it's a two-way conversation rather than a series of declarative statements. Sometimes all your neighbors need is a chance to air their concerns.

A friend of mine who also installs meadows gives her clients a small stack of brochures to share with neighbors. But even a typed-up piece of paper or email can be enough to make the case for a meadow. When sharing information, keep it simple and include details on why you're doing this. If you think your neighbors would be receptive, include some facts about the adverse effect lawns have on the environment, but keep it light so people don't get defensive. Or just skip bringing up lawns entirely, and focus on the benefits of adding a meadow to the neighborhood ecosystem.

Let your neighbors know what to expect, including information about timing and how long the meadow will take to fill out, plus how you plan to maintain it. You might let them know which flowers you're choosing and when they'll bloom. (Who doesn't like flowers?) Anticipate, and answer, a few questions they might have, like "what about ticks?"—a frequently-asked question if ever there was one.

One man I worked with headed off problems by putting up a sign that said *Meadow in Progress*. He'd printed and laminated it at a local copy shop after noticing several neighbors stopping and staring at his newly tilled yard. The sign led to some thoughtful exchanges with his neighbors, especially after he walked out to the curb to meet them. He told me the meadow kindled a few new friendships.

Change the rules. If you live in a town or neighborhood that prohibits meadow-keeping, you'll have to aim a little higher to create change. Outreach, education, and organizing is the *only* way to amend local ordinances. If you do your research and can build a sound ecological case for your meadow, you may be able to apply for a variance that allows you to convert your lawn, or even change your local laws.

BECOME A MEADOW ACTIVIST

I was saddened to read recently that the western monarch butterfly population is precariously low. Every fall as temperatures drop, all the monarchs on the western side of the Rocky Mountains migrate to the coastal mountains of California, where they cluster on trees in a few very specific locations over the winter. Their numbers have fallen below thirty thousand, which entomologists consider to be a critical threshold and represents a mere one percent of the population just thirty years ago! The population drop is just one data point among many that shows the urgency of the pesticide problem and, of course, the climate crisis. For me, it was also another call to action.

I invited a few local nurseries to participate in a trial using the Community-Supported Agriculture (CSA) model. But instead of buying into boxes of fresh greens and turnips, members pay for perennial plants at less than retail cost. In doing so, CSA members provide the financial support that gives growers a predictable revenue stream (typically in the spring, when they need capital the most). What I like about this concept is that as nurseries promote this opportunity, they're also educating the public about the ecological value of planting perennials for pollinators.

Where I live in western Massachusetts, I've watched a number of people and organizations step into the role of meadow activist. The founders of two of the nurseries participating in the CSA trial jumped into the business of promoting pollinator habitats with no prior experience. Between the two of them, *A Wing and A Prayer*, in Cummington, and *That's a Plenty Farm*, in Hadley, they've grown many thousands of pollinator-friendly meadow plants that have been planted in hundreds of yards.

Starting a nursery is more than most of us want to take on, of course. Another approach is to carve out one small area or a single project to focus on. A few years ago, Peggy MacLeod, a friend who lives in Northampton, just down the road from me, did exactly that. She had discovered that most perennial plant nurseries regularly use pesticides, including neonicotinoids. This is a nasty problem because when the treated plants flower, they attract—and then poison—local pollinators.

With this in mind, Peggy began contacting local nurseries and asking them to take a pledge to sell organically grown plants. A number of them balked, because a dirty secret in the nursery trade is that very few commercially grown plants are organic. But with a little persistence on Peggy's part, many committed to buying organic plants when possible. A few went even further and started requesting organic plants from suppliers who didn't yet offer any. Now, thanks to

Peggy's one-woman campaign, it's become a lot easier to find organic perennials for sale locally.

A few years ago when Peggy retired, she decided to put her impressive energy toward co-founding the Western Mass Pollinator Networks (WMPN)—pointing to yet another way to become a meadow activist. Four thousand or so different bee species live in the United States. Many are small, less than a centimeter long, and can fly only a few hundred feet at a time, so as their habitat shrinks, they grow increasingly isolated and threatened. But with the help of meadows and gardens that crisscross a city or a suburban area, these little insects, along with other pollinators, can find enough pollen and nectar to live on. In founding the WMPN, Peggy joined the hundreds of other citizen gardeners working around the country to create these pollinator corridors.

One prominent example can be found in Seattle, where a designer named Sarah Bergmann moved in 2007, not long after losing her mother. She resolved to support pollinators in part to deal with her grief. Her plan was to connect two existing green spaces, the University of Washington campus and a small plot of urban woods one mile away.

Sarah started knocking on doors and asking homeowners to grant access to the bit of grass between their sidewalks and the street—the area that the West Coast calls a "parking strip." She took on all aspects of the job, including navigating the bureaucracy that comes with getting approval for public projects. And, one after another, homeowners turned over their unused turf.

Sarah was eventually joined by more than three thousand university students and city residents who volunteered to convert these former parking strips into a narrow meadow, one mile long and twelve feet wide. The project was so successful, a second one-and-a-half-mile-long pollinator corridor is now being planned.

Sarah started a citywide movement, which is inspiring. But if you're not up for leading one of your own, you can join one of the organizations already devoted to meadow building. Many could use your help, from volunteering your time to providing financial support.

Xerces Society. This international nonprofit is one of my favorites. It supports the conservation of pollinators and other endangered species and their habitats, and works to reduce the use of pesticides. Every year, the Xerces Society organizes the western monarch count, in which volunteers gather data on the status of the declining monarch population.

Wild Ones. This group has chapters in eighteen states and promotes natural landscaping, including meadows. Its motto, "healing the earth one yard at a time," sums up its mission to influence and educate people about the importance of creating regenerative landscapes on their own properties. Wild Ones offers a way to get involved locally without having to start a movement from scratch.

Local organizations. A handful of regional and state organizations around the country are working to promote meadows and native pollinator habitats in their communities. In Texas, the Lady Bird Johnson Wildflower Center encourages people to collect milkweed seeds and replant them to provide habitat and food for monarchs. The Ohio Pollinator Habitat Initiative started a similar project in 2015, and since then Ohioans have gathered about 5,000 gallons of common milkweed seed pods, totaling more than 22 million seeds, to be redistributed throughout the state. The Florida Wildflower Foundation works to protect the state's local ecology by fighting to save local native wildflower habitats, as well as planting new ones.

You might have to do some research to find the existing groups and organizations in your own area but many others like these, including dozens of local gardening clubs, are working around the country to preserve and restore meadows and protect pollinators and their habitats.

IT'S UP TO US

Meadows offer a unique opportunity to help the planet from our own yards. They support many of the wild things that keep ecosystems healthy and store carbon, too, all without asking for much in return. Grasslands, natural as well as man-made, are also the ideal landscape for our climate-challenged times, thanks to their inherent resilience. If just a fraction of the existing lawns in this country were turned into meadows, the ecological impact, especially on threatened pollinator species, would be immediately significant.

Replacing unused lawns with meadows is honestly not that hard to do. Even if you don't have a yard, you can plant a few pollinator-friendly meadow plugs in a community garden or at a local school. Or sow a mix of native meadow seeds into an abandoned city lot. The ecology of the modern world is so fragmented by highways, urban areas, and monocultures like lawns and farms, all of which are biological deserts, that even one small meadow can be a haven for native insects and birds, and serve as a carbon sink.

Part of establishing a meadow is also a lesson in letting go. Basic maintenance aside, once you've designed and planted your meadow, your primary job is to give it the space and time it needs to reveal its own character. You can plan for every contingency with your site prep and design, but nature will have the last word. And if you've honored that fact by working with your environment, your meadow will very likely succeed, and be magnificent.

Imagine yourself looking out on a meadow—your own or one you've helped build. Birds snatch up seeds for one last meal as the sun sets and the sky dims, leaving a soft glow on the horizon. As the light fades, the sound of crickets emerges and the air begins to feel cool and damp. You breathe it all in—the sweet, green smell of grasses and the floral aroma of blooming wildflowers. A bat swoops by almost imperceptibly, barely topping the meadow plants on a hunt for mosquitos and small insects. As the chorus of crickets grows louder, a toad emits its prehistoric-sounding trill. The chill in the air sharpens, and you think about going inside, but you linger just one more moment, mesmerized by this beautiful show. When I envision the future of our planet, more of *this* is what I'd like to see.

QUESTIONS

What about ticks?

Ticks used to be confined to the Northeast, but thanks in part to warming winters this bearer of so many diseases, Lyme among them, has taken up residence in almost every state. Ticks like to drop on their prey from tall plants and grasses, which is one big reason many people tell me they have no interest in giving up their cropped lawns. But with a little planning, you can significantly minimize the risk of being ambushed.

- Mow pathways wide enough for people and dogs to walk through without brushing against any grasses or plants.
- Stay out of your meadow during the warmest part of the summer, when ticks are most likely to jump you. Since an established meadow requires pretty much zero maintenance (other than mowing), this shouldn't be too hard.
- If you must walk into your meadow during tick season, wear light-colored clothing so they're easy to see, tuck your pants into your socks, and use a nontoxic repellent. Wash your clothing afterward and dry it at a high heat, which kills any ticks that might be hiding there.
- Perform daily tick checks.

Meadows sound pretty rugged. Is it possible to grow one anywhere?

Meadows adapt well to conditions that most plants find trying, and can do quite well in poor, beat-up soil. You can find meadow plants in abandoned quarries, along busy roadsides, and sprouting out of rocks and concrete. Their needs are minimal—a half day of sunshine and at least twenty to twenty-five inches of annual rainfall.

Meadow plants can grow in very arid zones, as long as they're dry-loving native perennials. In places like California, where residents are encouraged to get rid of their lawns for conservation reasons, these dry meadows offer an excellent solution. Meadow plants don't do as well in locations that are overly moist or drain very slowly. But even then, it's possible to get good results if you match them to your conditions.

Bottom line: Meadows do best in areas that are neither too wet nor too dry. If you live in an area that receives fewer than twenty inches of rain a year, you probably can't count on your meadow lasting all year long.

Is it okay to buy seed blends rather than mix my own?

Premixed seed packets are fine if the plants are native and the mix tailored to your climate and site conditions. But your plant choices will be limited and you won't get the pleasure of putting together your favorite colors and plant preferences, or closely matching them to your site.

If you go the premixed route, avoid mixes that contain lots of annuals or non-natives. Prairie Moon Nursery is one example of a supplier that offers a nice variety of seed mixes.

Are there benefits to planting annual flowers in my meadow?

Annual meadow seed mixes are among the reasons a wave of meadow fever burned out toward the end of the twentieth century. They'd blossom dramatically in the first year and then disappear, leading to the misperception that meadows don't last.

But an upside of annuals is that they grow quickly. If you want to inject near-instant beauty into your meadow and feed pollinators while waiting for your perennials to grow in, go ahead!

To leave enough room for your perennials, plant slightly less of your nurse crop and use thin, wispy annuals, such as poppies, cosmos, or love-in-the-mist.

Nothing is growing in my soil. What can I do to bring it back to life?

Meadow plants can thrive in almost any type of soil. But if your lawn has lots of bare spots where not even weeds can grow, you might be looking at dead soil.

Lifeless soil can be a casualty of a lawn treated with chemicals. To jump-start microbial soil life, consider tilling, which often works, especially if there's enough vegetation to turn under. If there isn't, plant a cover crop like clover or buckwheat, which are hardy enough to grow even in depleted soil.

Once they've grown three or four inches high, till them back under—your gift to the soil. As cover crops decompose they add nitrogen and phosphorus to the soil, along with plant matter that feeds beneficial microbes and mycorrhizae and makes it easier for the soil to conserve moisture.

The next step is pretty straightforward: Go ahead and plant a fresh crop of perennials.

I tried to plant a meadow into my lawn before reading your book, and it's still weedy and grassy. What do you recommend?

Before buying any seeds or plugs, I recommend going through a site assessment to make sure you're matching your plants correctly. Then mow your grass and meadow plants close to the ground, and sprinkle in more seed or add new plugs. Seeding can be done any time of year, though spring and fall is better than summer. But I recommend planting new plugs into existing lawn in the fall, when they're less likely to be taken over by neighboring plants.

If your meadow is *really* weedy, you may need to mow it more than once during the first growing season and selectively pull out or cut back the worst of the weeds. But if you've tried that for more than a year or two and still don't see many meadow plants, I recommend tilling it under and starting over.

Making this call depends to some extent on what types of weeds are growing in your yard. A site with lots of persistent perennial weeds, like bittersweet or poison ivy, is more likely to require a rebuild than one with mostly annual weeds.

My meadow is still thin more than a year later. Should I add new plants?

First, try to understand why your seeds or plugs aren't taking. Plugs are easier to figure out. Perhaps mulch covered the base of the plugs and they became moldy and died. Or, it rained too much and the roots rotted. Maybe it was hot right after you planted and you forgot to water them, so a number of plants crisped.

Finding out why seeds aren't taking can be trickier because they're almost invisible, and present fewer clues. If this happens I always walk back through the entire assessment and design process to make sure I didn't miss anything.

If you've determined that the way you planted the meadow was the problem, go ahead and replace any missing plugs, or sprinkle in more seed following the instructions I've shared in this book. If, instead, you think the plants you chose aren't a good match, find ones that are. You can add your new plants to what you've already created, or clear your site and start from scratch.

What if invasive or aggressive species appear in my meadow?

Most weeds can be managed with regular maintenance. But some species, like Japanese knotweed or kudzu, are so aggressive you'll want to control these intruders immediately after noticing them. Dig out any invasive species as best you can, and keep cutting back anything that returns.

Be vigilant. You want to prevent invasives from unfurling new leaves, since leaves equal food. By cutting back new growth, you're depriving them of energy and, eventually, they will weaken and die. If a big patch of these invasives develops, consider solarizing to get rid of them.

How do I do a controlled burn?

The stakes are high with an annual burn, and timing is critical. You want to avoid burning when your meadow is actively growing. Early spring is the best time because old meadow vegetation is dry and burns easily and new vegetation is not yet above ground. As with mowing, waiting until spring also allows a meadow to safely harbor overwintering insects and their eggs.

To burn a meadow safely, make sure the entire perimeter can act as a firebreak. A firebreak is an area free of burnable material and wide enough that flames can't leap across it. A mowed strip ten to twelve feet wide can work as a firebreak around most meadows, but the taller the meadow grasses, the wider it needs to be. The perimeter can also be secured using something called back burns, which are small, carefully controlled burns that are either allowed to burn out or extinguished by flails or hoses.

Once a perimeter has been cleared, pick a day that is dry—wet meadows won't burn—and not too windy. A very light breeze can be helpful but anything more than that is dangerous. Start your burn by lighting fires in a line so the flames burn as a front across the meadow. Allow and encourage it to burn across the entire meadow until it reaches the farthest firebreak, where it will burn itself out. This can work on small sites and large sites too, as long as they're very carefully controlled and not near anything flammable.

What else can I do to support meadows?

Consider eating more plants, if you aren't already doing it. The growing global demand for meat is threatening the very existence of grasslands, given how much land it takes to grow food to feed livestock. The most recent report from the U.N.'s Intergovernmental Panel on Climate Change found that 30 percent of cropland is used to grow grain for animal feed.

Clearing forests and converting grasslands into pasture and cropland also releases carbon into the atmosphere, and depletes the soil. Since people began farming, the world's cultivated soils have lost between 50 and 70 percent of their original carbon stores.

The panel concluded that any meaningful climate strategy has to include a concerted effort to restore organic matter to grassland soils, reduce erosion, and move to a more plant-based diet. If you care enough to convert your lawn into meadow, it makes sense to reduce your footprint elsewhere when you can.

SEED SUPPLIERS

The following seed companies are all committed to selling native seeds. Since they're doing this work in part to support regenerative landscapes, they tend to be meticulous about the quality of their seeds. I've included a sampling of suppliers from across the country so you can work with the one closest to you. That said, many of these seed houses also serve more than their local communities.

Ernst Conservation Seeds (Pennsylvania)
Florida Wildflowers Growers Cooperative (Florida)
Heritage Seedlings & Liners (Oregon)
Larner Seeds (California)
Michigan Wildflower Farm (Michigan)
Missouri Wildflowers Nursery (Missouri)
Native American Seed (Texas)
Native Wildflowers and Seeds from Ion Exchange (Iowa)
Prairie Moon Nursery (Minnesota)
Prairie Nursery (Wisconsin)
Roundstone Native Seed (Kentucky)
Toadshade Wildflower Farm (New Jersey)
Western Native Seed (Colorado)

ACKNOWLEDGMENTS

It's an honor to have the opportunity to write this book. If we want to live on a truly healthy planet, we need to address the effect lawns have on the environment. For decades I've been exploring how to facilitate regeneration to help make that happen.The perspectives and information I've shared on these pages are the fruits of that research. My hope is that meadows become a larger part of the conversation on how to be better stewards of the earth.

This book would not have been possible without the vision of Clare Ellis, who understood why it needed to exist. Her steady hand also made the entire writing process much more manageable despite the small window of time I had to work with. As an editor she honored my voice and experience and worked tirelessly to elevate them, helping me achieve a much better book, and for this I am grateful.

The careful editing support I received from Christine McKnight also made it possible to pull this book together. Her meticulous eye and attention to detail were invaluable. Similarly, it was a pleasure to have the support of illustrator Kristen Thompson. Kristen beautifully expressed the character of each plant I profile here, and her other illustrations support the book hand in glove.

I'm also grateful for my trusted readers, kdb Dominguez, Virginia Sullivan, Ted Watt, and Baron Wormser, who combed through the book in record time and gave me invaluable feedback. Thank you, too, to Sophie Jones for their help with the plant profiles.

And lastly, gratitude to my parents for following their hearts and doing what felt right to them. Growing up the way I did gave me a running start when it comes to connecting with and understanding nature, and in the long run, helped make this book possible.

NOTES

THE GENEROSITY OF MEADOWS

a closely cropped area of turf that totals more than 63,000 square miles... lawns the biggest irrigated crop grown in the United States. "A Strategy for Mapping and Modeling the Ecological Effects of US Lawns," *The International Society for Photogrammetry and Remote Sensing Volume XXXVI (2015)*. https://www.isprs.org/proceedings/XXXVI/8-W27/milesi.pdf.

over forty million acres of land in the continental US were found to have some form of lawn on it... "Mapping and Modeling the Biogeochemical Cycling of Turf Grasses in the United States," *Environmental Management Journal* 36 (2005). https://link.springer.com/article/10.1007/s00267-004-0316-2.

landscape irrigation is estimated to account for nearly one-third of all residential water use, totaling nearly nine billion gallons per day. "Outdoor Water Use in the United States," Watersense/EPA https://19january2017snapshot.epa.gov/www3/watersense/pubs/outdoor.html.

LAWN TROUBLE

Americans use 100 million tons of fertilizer on their lawns each year. Botanic Gardens Native Lawns, Cornell University Sustainable Campus. https://sustainablecampus.cornell.edu/campus-initiatives/land-water/sustainable-landscapes-trail/botanic-gardens-native-lawn

For every ton of fertilizers manufactured, two tons of carbon dioxide are produced. "Energy Consumption and Greenhouse Gas Emissions in Fertilizer Production," International Fertilizer Association.

More fertilizer is applied than plants can absorb..."Fertilizer is bad stuff, and not just because it can blow up your town," Tom Laskaway, *Grist*, April 2013.

Increased fertilizer use over the past fifty years is responsible for a dramatic rise in atmospheric nitrous oxide."Fertilizer use responsible for increase in nitrous oxide in atmosphere," Kristie Boering, UC Berkeley, *Nature Geoscience*, April 1 2012.

Conventional agriculture is infamous for producing enormous aquatic dead zones from high levels of chemical runoff. "How Fertilizers Harm Earth More Than Help Your Lawn: Chemical Runoff from Residential and Farm Products Affects Rivers, Streams and Even the Ocean," *Scientific American*, July 20, 2009.

Homeowners use up to ten times more chemical pesticides per acre than farmers do. "What's happening to the frogs?" US Fish & Wildlife Service, 2006.

The agency can "conditionally approve" pesticides based on data provided solely by the manufacturer. "NRDC Report: More than 10,000 Pesticides Approved by Flawed EPA Process," Natural Resources Defense Council (NRDC), March 2002.

The agency continues to greenlight chemicals before a full evaluation of health and environmental impacts has been completed... "Ban Dangerous Pesticides," NRDC, https://www.nrdc.org/issues/ban-dangerous-pesticides

though it has said it's working to make testing more rigorous. "Conditional Pesticide Registration," EPA, https://www.epa.gov/pesticide-registration /conditional-pesticide-registration.

approved its use before fully understanding the terrible impact it would have on monarch butterflies, bees, and other birds. "The environmental risks of neonicotinoid pesticides: a review of the evidence post 2013," Thomas James Woods, *Environ Sci Pollut Res Int.* 2017; 24(21): 17285–17325. "EPA: Neonicotinoid Pesticides Pose Serious Risks to Birds, Aquatic Life," Center for Biological Diversity, *Ecowatch*, December 2017.

the total biomass of insects worldwide has plummeted by a precipitous 80 percent. "Plummeting Insect Numbers 'Threaten Collapse of Nature,'" Damian Carrington, *Guardian*, February 10, 2019. "Insect 'Apocalypse' in U.S. Driven by 50x Increase in Toxic Pesticides," Stephen Leahy, *National Geographic*, August 6, 2019.

the populations of more than 75 percent of songbirds and other birds that rely on agricultural habitat have also significantly declined..."Huge Decline in Songbirds Linked to Common Insecticide," Stephen Leahy, *National Geographic*, September 12, 2019.

sufficient evidence of carcinogenicity to classify glyphosate in 2015 as "probably carcinogenic to humans." *IARC Monograph on Glyphosate*, World Health Organization, International Agency for Research on Cancer, March 2015.

180 million pounds of it were being used nationally. "Pesticides Industry Sales and Usage: 2008-2012 Market Estimates," Donald Atwood and Claire Paisley-Jones, EPA, 2017.

Glyphosate was found in 86 percent of air samples and 77 percent of rain samples... "Pesticides in Mississippi Air and Rain: a Comparison between 1995 and 2007," Michael S. Majewski et al., *Environmental Toxicology and Chemistry*, Volume 33, Issue 6 (February 2014): 1283–1293.

In 2016 alone, more than 286 million pounds of it were used. "Pesticides," Ohio-Kentucky-Indiana Water Research Center, United States Geological

Survey. https://www.usgs.gov/centers/oki-water/science/pesticides?qt-science_center_objects=0#qt-science_center_objects.

Lawn chemicals also account for the majority of wildlife poisonings reported to the EPA. Lawn Pesticides, NY Audubon Society https://ny.audubon.org/conservation/lawn-pesticides

Pets kept in yards treated with herbicides and other chemicals experience significantly higher occurrences of malignant lymphoma—70 percent higher—than pets in organic yards. "Household Chemical Exposures and the Risk of Canine Malignant Lymphoma, a Model for Human Non-Hodgkin's Lymphoma," Biki B. Takashima-Uebelhoer et al., *Environmental Research* Volume 112 (January 2012): pages 171–176.

repeated use can degrade soil biology... "The Impact of Glyphosate on Soil Health:. The Evidence to Date," Soil Association. https://soilassociation.org/media/7202/glyphosate-and-soil-health-full-report.pdf.

Seventeen million gallons of gasoline are spilled every year in this country just while filling lawn mowers. "How to Pick a Lawnmower that's Easy on Man and Nature," *Scientific American*, June 2008.

Lawn mowers emit ten times more hydrocarbons than a typical car for every hour of operation. "Native Plants for Wildlife Habitat and Conservation Landscaping: Chesapeake Bay Watershed," Britt E. Slattery, Kathryn Reshetiloff, and Susan M. Zwicker, US Fish & Wildlife Service, Chesapeake Bay Field Office, Annapolis, MD, 2003.

Spews as much smog-forming pollution as driving a 2017 Toyota Camry for 300 miles. "Small Engines in California," California Air Resources Board, August, 9, 2017, https://ww2.arb.ca.gov/resources/fact-sheets/small-engines-california.

The amount of emissions released is the equivalent of driving a car for one hundred miles. "Zero-Emission Landscaping Equipment," California Air Resources Board, https://ww2.arb.ca.gov/our-work/programs/zero-emission-landscaping-equipment.

four times greater than the amount of carbon stored by grass. "Turfgrass management may create more greenhouse gas than plants remove from atmosphere," Amy Townsend-Small, *Geophysical Research Letters*. https://news.uci.edu/2010/01/19/turfgrass/

The average homeowner will spend 150 hours a year tending to their grass. *American Green: The Obsessive Quest for the Perfect Lawn*, Ted Steinberg, (New York: W.W. Norton & Company), 2006.

Collectively we spend forty billion dollars on lawns annually. "Blades of Glory: America's Love Affair with Lawn," *The Week*, June 24, 2011.

Sales would hit 4.2 million—and they haven't slowed since. *American Green: The Obsessive Quest for the Perfect Lawn*, Ted Steinberg, (New York: W.W. Norton & Company, 2006), 10–13, 26, 33.

well over 5 million gas powered mowers are still sold in the US every year. "Cleaner Air: Gas mower pollution facts," *People powered machines*. https://www.peoplepoweredmachines.com/faq-environment.htm

REGENERATIVE SCAPES

About half of all carbon released into the atmosphere every year is absorbed by the planet's oceans, plants, and soil. "Earth Still Absorbing about Half Carbon Dioxide Emissions Produced by People: Study," NOAA Headquarters, Phys.org, August 1, 2012.

The research on grasslands finds they're a cost-effective and scalable solution for carbon absorption. "Grasslands Among the Best Landscapes to Curb Climate Change," Kelly April Tyrrell, *UW MadScience*, November 15, 2018.

Meadows perform as well as or better than a forest when it comes to sequestering carbon. "In Wildfire-Prone California, Grasslands a Less Vulnerable Carbon Offset Than Forests," Kat Kerlin, *UC Davis Science & Climate*, July 9, 2018.

North American grasslands can store between four and a half and forty tons of carbon per acre–and that's just in the top twenty centimeters of soil. Chapter 10: Grasslands in *Second State of the Carbon Cycle Report (SOCCR2): A Sustained Assessment Report*, ed. N. E. Pendall et al. Washington, DC: U.S. Global Change Research Program, 2018.

The soil under two and a half acres of healthy prairie was found to have absorbed as much carbon as what 150 cars emit over the course of a year... "Links Between Grasslands and Carbon Storage," Alberta Prairie Conservation Forum, September 13, 2011.

An established meadow is able to store 70 percent more carbon than a monocrop, like turfgrass. "Soil Carbon Sequestration Accelerated by Restoration of Grassland Biodiversity," Yi Yang et al., *Nature Communications*, February, 12, 2019.

COMMUNITY BUILDING

Senator Rand Paul had six ribs broken when he was attacked by a neighbor... "Court Documents Give New Details about the Yard Dispute that Left Rand Paul with 6 Broken Ribs," Jen Kirby, *Vox*, June, 12, 2018.

The Western monarch butterfly population is precariously low. "Western Monarch Butterfly Numbers Critically Low for Second Straight Year," Liz Kimbrough, *Mongabay*, February 26, 2020.

Four thousand or so different bee species live in the United States. "The Buzz on Native Bees," US Geological Service, June 15, 2015.

Many are small, less than a centimeter long, and can fly only a few hundred feet at a time. "Tiny Pollinators Need Wildlife Corridors Too," Michelle Nijhuis, *Atlantic*, January 19, 2017.

QUESTIONS

30 percent of cropland is used to grow grain for animal feed... "Climate Change and Land," Intergovernmental Panel on Climate Change, 2019.

On average, cultivated soils have already lost between 50 and 70 percent of their carbon stores. "Soil as Carbon Storehouse: New Weapon in Climate Fight?" Judith D. Schwartz, Yale Environment 360, March 4, 2014.

BIOS

Owen Wormser was born and raised off the grid in rural Maine. He received a degree in landscape architecture in 1998 and moved to western Massachusetts to start a landscape design and installation service. Since then he has built hundreds of regenerative landscapes influenced by his study of horticulture, permaculture, organic agriculture, and ecology. In 2010, Owen started Abound Design, which provides design and installation services with a focus on creating sustainability, regeneration, and beauty. He also runs a nonprofit that provides educational resources and hosts workshops on regenerative growing.

Kristen Thompson is an illustrator, gardener, environmentalist, and writer based in New Jersey. She has been taking art classes since middle school and studied at the Parsons Pre-College Academy and the Putney Summer Arts Program. While at Marlboro College, she studied edible, perennial landscaping and designed a garden for the college farm, along with an illustrated field guide to the native plants she used. Kristen has also written and illustrated an environmental fantasy novella.

More citizen gardening books...

A must-have resource for home gardeners looking to take their conservation efforts to the next level. With hard-earned knowledge and conversational clarity, Tucker demystifies the concepts of regenerative agriculture, translates them to the garden level, and guides the reader both philosophically and practically.

Stephanie Anderson,
***author of* One Size Fits None**

Great for new and experienced gardeners, *Growing Perennial Foods* is worth the purchase for the recipes alone.

Gardening Products Review

Not only does Acadia know what she's talking about, she is passionate about it.

Trish Whitinger, National Gardening Association

Beautifully written and illustrated, this book will be a well-thumbed addition to your gardening library.

The Northern Light

Acadia Tucker believes that taking cues from how plants grow in the wild will allow for cultivated gardens that produce bountiful harvests while addressing concerns about global climate change. Her guide moves through all the steps needed to create a healthy, nurturing bed.

Anne Heidemann, American Library Association

Acadia Tucker is on a mission to get more of us thinking about the power of regenerative agriculture. By the end of the book, you'll feel inspired enough to start your own Climate Victory Garden.

Jes Walton, Green America

Tucker helps us tap into the deeper meaning of gardening *and* grow good food at the same time.

***Anne Biklé, coauthor,* The Hidden Half of Nature: The Microbial Roots of Life and Health**

... from Stone Pier Press

Growing Good Food is a solid introduction to the larger conversation of how to make a difference on this planet with one's own land. I recommend it to every gardener.

Peggy Riccio, PegPlant

My father, Wendell Berry, says that this kind of work is radical now, when public attention is focused on global solutions. This work is what people are for.

Mary Berry, Executive Director, The Berry Center

In this well-informed call to action, Acadia Tucker sets out to simplify regenerative gardening so anyone can do it. She succeeds! With step-by-step instructions and chapters dedicated to your favorite veggies, readers will be inspired to grow food and save the planet, all from the comfort of your backyard.

Jes Walton, Green America

Tucker has a unique perspective on climate change. Working as a farmer from Washington to New Hampshire, she has seen radical shifts in climate that decimated sensitive annual crops but spared perennials. She has also seen the difference it makes to those crops when soil contains an abundance of organic material. It's a fascinating read.

Todd Heft, BBOG

I love this book. *Growing Good Food* is great for beginner gardeners who care about the climate.

Lucy Biggers, Now This Media

Acadia Tucker's book is an important read for any backyard grower who wants to make a positive impact on the climate in their own patch of dirt.

Robyn Rosenfeldt, Pip Magazine

I've been inspired to start sheet mulching my entire yard and now think about ways to help keep carbon in the ground longer.

Scott Vanderlip, Slow Food

Acadia Tucker's new book shows what it takes for beginners to throw themselves into regenerative agriculture.

Lindsay Campbell, Modern Farmer